List of Figures

List of Tables

Chapter 1

Getting Design Right

What Do We Mean by "Design"?

To some people, the word *design* means graphical or artistic design and is therefore in the domain of the artist. To other people, it means engineering design and is therefore in the domain of the engineer. If you are neither an artist nor an engineer, it would seem that you have little to do with design. But, if you broaden your definition of design to include planning and general problem solving, then you will see that you are engaging in design on a daily basis. Design is the creative part of your day, as opposed to the routine part such as brushing your teeth. When you enter a kitchen and decide what you are going to make for dinner, you are engaged in design. When you are composing a speech in your head that you will be delivering to a community group, you are engaged in design. Rearranging furniture, applying for jobs, and planning vacations are all design activities in the broad sense of the word. *Design* is the activity of imagining the future, setting goals, and finding ways to achieve those goals.

This is a practical book about design. To be practical, we focus on product design problems. That is, we imagine that you want to make something or hire someone to make it. You want to be involved deeply in the design process to make sure you get what you want. The "something" you want can be practically anything. It could be a consumer product such as a household appliance; it could be a software application or a business Web site. As a result, this book will have the most immediate interest to you if you are engaged in a product development process. But it has broader applicability. Many of the principles and techniques in this book are generally applicable to all design problems. They can be used to design human organizations: the mission of the organization, the roles of different individuals in the organization, the rules governing their interaction, the flow of work, and the command and control. You could thus use these techniques to design both the product and the business that will make money from the product. Though we focus on product design, you should recognize that by studying this book you are learning a general approach to problem solving.

In this book, we also lay the foundation for the design of complex systems: integrated networks of hardware, software, and human organizations. As an example of a complex system, think of an air traffic control system in which airplane pilots and ground controllers work together using radar, computer, and communication systems to avoid aircraft collisions and to manage the sequence of takeoffs and landings at a busy airport. Perhaps that seems out of your grasp. We do

1

not address the technological details of such complex systems, but we do show how the complexity of such systems is managed. It is handled with a straightforward design process applied to layer after layer of detail. The design process we describe in this book works whether you are designing a child's toy or an air traffic control system, a new line of clothing or a spacecraft.

Why "Getting Design Right"?

Every job you do involves design. Perhaps the design is well established, and the job has become routine. You no longer think about objectives, materials, sequence, and technique; you simply perform the job in a mechanical fashion. But, once the job was new and you were either taught how to do it or you figured it out on your own. If you figured it out on your own, you were the designer. Even if the job was as simple as finding out how to get to work, we are, all of us, designers in this sense. Design is the creative part of any job: identifying the objective, exploring the possibilities, picking a course of action, and testing and improving until we get it right.

Getting design right—fitting the result to the need—is greatly satisfying. In particular, if you design a product that pleases everyone who comes into contact with it, you can experience great joy as a designer. If you do it with economy, using the least effort and purest form, then your work has grace and elegance. Getting design right can also be commercially rewarding. The invention of the MP3 player, a portable digital audio player, was a breakthrough. The design and marketing of the iPod, Apple's version of an MP3 player, was a triumph. Personal satisfaction is therefore one motivation, and commercial success, another. You can have both.

Getting design wrong is a waste. It is a waste of time, of money, and of creative energy. We have all struggled with products that seem to work against us: devices that are too heavy, too complicated, or too flimsy. We have all experienced processes that are inefficient or unreliable: waiting in line for one agent, who then tells us we are in the wrong line; multistep processes to perform an operation so common we think it should be only a single mouse-click; and car repairs that seem to last just longer than the repair warranty. If you can think of an example of consumer frustration—of something that has wasted your time or money—then you have likely identified some form of design flaw. The extent of consumer frustration you have experienced is an indication that getting design right is not as easy as you might imagine. Getting design right is a process of avoiding design flaws to the best of your ability.

What Can Go Wrong?

We have collected many stories about design, both failures and successes, to illustrate the importance of getting design right. Each chapter in this book features at least one example of a design failure. Each failure can be traced, in part, to a failure to follow the process of getting design right properly. There are also stories of design successes. All of these stories can be read independently of the main text. Discussion questions at the end of each chapter are meant to stimulate your thinking about the stories.

The Sinking of the *Vasa*

In 1625, King Gustavus Adolphus of Sweden commissioned construction of a new flagship, the *Vasa* (Figure 1.1). Sweden needed a strong navy to dominate the Baltic Sea.

Figure 1.1   Restored seventeenth-century warship, the *Vasa*, Stockholm, Sweden. (Courtesy of Staale Sannerud.)

The *Vasa* was intended to be the world's largest warship, with two gundecks and 64 guns. Hendrick Hybertszoon, a master shipwright from Holland, was selected to build it. Initially, there were no written requirements from the king. The master shipwright assumed the ship would be 108 feet in length, but, after the first review, King Gustav insisted on a 120-foot ship.

Having a second gundeck meant adding 50 brass 24-lb cannons (at one ton each) to an already altered design. The master shipwright estimated that another 130 tons of rock ballast would be required under the deck to counterbalance the new cannons. However, there was no room for the additional ballast so it was left out.

The king was eager to show off the new flagship to the world. He pressured the shipwrights to complete the ship several months ahead of schedule. The master shipwright, Hybertszoon, died a year before the *Vasa* was finished. His less-experienced brother, Arendt de Groot, completed the project.

Stability tests were conducted. These involved having 30 sailors run from one side of the ship to the other. The ship appeared to be unstable: It almost capsized on the third run. Klas Fleming, the navy admiral present at the test shrugged off the problem.

On a Sunday in August 1628, the *Vasa* set sail before a large audience that included foreign diplomats whom the king hoped to impress. After a short distance, and in full view of the shocked audience, a gust of wind caught the mainsail, the ship heeled over, and water began to pour in through the gunports. The gunports had been left open even though this was the maiden voyage of a ship with known stability problems. The ship capsized and immediately sank. At least 30 of the 150 people on board drowned.

The story of the *Vasa* is popular in design circles (Love, 1993) because it illustrates the problems that continue to plague many projects, both large and small, even today: incomplete and shifting requirements, lack of meaningful trade-off analysis, accelerated construction timetables, leadership failures, and inadequate test and review cycles, to name a few.

What Is There to Learn?

Getting design right is a skill. There are principles involved. Perhaps you already use some of these principles, though you use them unconsciously. Product successes can be traced to the consistent, intelligent use of these principles, and failures can be traced to their misapplication. There are also techniques developed by good designers that help them to implement these principles. Furthermore, there is a process for design: a place to start and steps to follow. This process of getting design right is applicable, with adaptation, in all design contexts. You may skip some steps and simplify others, but the basic process is sound. Getting design right, therefore, is something that can be taught and learned.

Why a Systems Approach?

We have qualified the title of this book with the phrase "a systems approach." Perhaps that is superfluous; any book that seeks to explain how to get designs right would necessarily have to adopt something like a systems approach. We use this phrase to emphasize that context is important. A successful race car design does not focus simply on speed: Strapping a rocket to the vehicle would maximize speed. Bad designs often do something extremely well. The failure comes in neglecting other aspects of the design that are important in the context of the product's use, such as the ability of a race car to maneuver around sharp corners at high speed. The phrase "a systems approach" suggests that you need to look at all aspects of the product in the context of its use within a larger system. There is a formal approach to studying an object in context—a systems approach.

A *system* is simply a collection of objects in relation to each other. A *systems approach* is therefore a procedural focus on relationships. Does your product vibrate in use? Taking a systems approach would lead you to explore the impact that vibration might have on the fasteners that hold your product together. Understanding relationships is critical to getting design right, and there are formal techniques to ensure that all relevant relationships in a design context are discovered and considered.

We also emphasize a systems approach because the language of systems is useful in all design contexts. Every discipline has evolved a language to describe the artifacts and methods within its design domain. Electrical engineers speak of circuits, mechanical engineers discuss forces, software engineers describe algorithms, and artists discuss medium and technique. The language of systems attempts to cut across disciplines and establish a more generic vocabulary and grammar with which to discuss design. We use abstractions such as entities, relationships, goals, functions, behavior, interfaces, and requirements to describe systems, and we become comfortable translating these concepts into specific artifacts and methods in different domains. The language of systems permits experts in different domains to talk with one another. It provides you with a universal language of design.

Finally, we use the phrase "a systems approach" because one of the techniques that is central to this approach is a technique known as "diving and surfacing." Diving refers to developing a detailed view of a system. Surfacing refers to abstracting from a detailed view to a more general view. It is the ability to surface, to abstract, and to "see the forest from the trees" that distinguishes the systems thinker from an individual who is trapped in discipline-specific details.

Design or Engineering?

This book is focused on product design and project management. It is not a text for detailed engineering. For the most part, we stop short of formulating and solving engineering problems. Instead, we focus on a general approach to design using generic techniques and expressing design in a systems language. It is a pre-engineering approach in the sense that it takes the design problem to the point where specific engineering problems can be posed and handed over to domain experts such as mechanical, electrical, or software engineers. But it is also a post-engineering approach in the sense that it identifies tests that the engineering solutions must satisfy. In short, it takes the perspectives of the product design owner and the project manager with the understanding that detailed design, build, and test can be subcontracted to other organizations.

Figure 1.2 illustrates a "Vee" diagram (Fossberg and Mooz 1992). A design project timeline moves from left to right in the figure, but the movement is downward into engineering detail on the left and then upward through testing on the right. The basic point is that each level of test should map back to an earlier design phase. In this text, our focus is not on the "engineering design–build–test" sequence below the line. Instead, in Getting Design Right we focus on the "pre-engineering design" and "post-engineering test" that occur before and after this "inner V." Ours is more of a systems view of design.

The examples and assignments in this book are incomplete because they stop short of creating true executable designs—that is, blueprints or drawings from which actual products can be constructed. The designs of this book are conceptual designs: product concepts thought out with as much detail as possible in a pre-engineering phase.

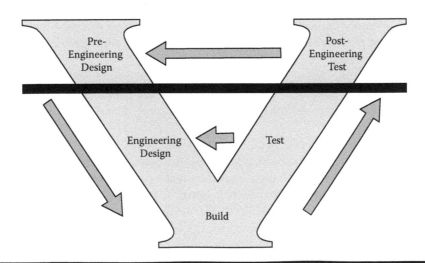

Figure 1.2   The "Vee" diagram: linking test to design.

In spite of the lack of traditional engineering content, this book is intended to be useful to engineers because the curricula of most engineering schools focus on analysis, not synthesis. As an engineering student, you are, or were, expected to discover the principles of design either on your own or on the job. The emphasis on design and synthesis is missing from many curricula. You may have been practicing engineering for many years without having been exposed to the formal approach presented here. If this is the case, you will readily see how to integrate these ideas into your daily engineering work. It will also help you to understand better your role in the overall product development process within your organization.

For Whom Is This Text Designed?

This book is written for a broad audience: students interested in the design–engineering interface, project managers and staff, and practicing engineers without formal design or project management training. No matter which category you fall into, you will find the material in this book accessible, provided that you have a reasonably good background in secondary school science and mathematics and you have used spreadsheet software. More advanced techniques of systems engineering and analysis can be found in graduate-level textbooks. We have excluded them from this book.

What Is the Design Process?

As mentioned earlier, there is a process for design: a place to start and steps to follow. This process of getting design right is applicable, with adaptation, in all design contexts. In this book, we use the name "The Eight Steps to Getting Design Right" to describe this process. The eight steps are as follow:

1. *Define:* Define the problem. In this step, you will identify the product concept, define a project to design the product, establish the system context of the product, capture the "voice of the customer," and identify functional requirements for the product.
2. *Measure:* Measure the need and set targets. In this step, you will take ordinary English statements of product and functional requirements and, from them, develop measurable performance targets.
3. *Explore:* Explore the design space. The product concept you started with may not be the best design idea to meet all of the requirements that you have identified. In this step, you will use brainstorming and other techniques to explore the range of design possibilities systematically.
4. *Optimize:* Optimize design choices. After exploring the design space, you will have a number of workable integrated concepts from which to choose. In this step, you will compare these integrated concepts with each other from many perspectives and then select the best one. You will also set design parameters to optimize trade-offs.
5. *Develop:* Develop the architecture. In this step, you will define the functional behaviors of the product in greater detail, identify the components or subsystems, define how these components will work together to achieve the product behaviors, and identify the physical constraints that each component must satisfy in order meet the product level targets.

6. *Validate:* Validate the design. In this step, you will develop a master test plan to ensure that the product will meet the needs for which it is being designed. You will also identify the risks of the project and develop a strategy for mitigating those risks.
7. *Execute:* Execute the design. In this step, you will develop a plan for bringing the project to a successful and timely conclusion.
8. *Iterate:* Iterate the design process. For simple design problems, the preceding steps should be sufficient. However, complex systems must be designed in layers. In this step, you will learn how this design process is adapted and repeated for the next layer of detailed design. This time, however, you will understand the need for information tools to manage the explosion of detail and to ensure that the complex product will meet the needs for which it is being designed. Iteration is also required whenever you hit a roadblock in the design; further, you can learn to close out every design cycle with a systems view of your accomplishments.

The Getting Design Right process is depicted as a cycle of activities in Figure 1.3. The Eight Steps to Getting Design Right also provide the structure for this book. We devote a chapter to each step in the process: a total of eight chapters beyond this introduction. Refer to the table of contents to see how each of these major steps is broken down into further steps.

Learn by Example

The language and concepts of this book can seem quite abstract unless you can see clearly how to apply them. To make the book practical, we apply the steps of the design process to a specific example. It is a simple example: the design of a child's toy. Our initial design criteria are that the toy should be fun, safe, attractive, and affordable. From this simple starting point, we apply the Eight Steps to Getting Design Right process and we end with a product development plan and a collection of product definition documents that are ready for detailed engineering, design, and test. Throughout the book, we define system concepts abstractly and then illustrate them using this central example.

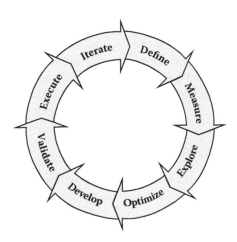

Figure 1.3 Getting design right cycle: Eight Steps to Getting Design Right.

Learn by Doing

This is not a book that you should attempt to read cover to cover without interruption. It is unlikely to have much impact upon the way you approach design unless you actually practice the steps. The techniques tend to build on each other, so you should master each step before advancing to the next. This book is intended to be part of a course of study in getting design right. In this course of study, you will apply the Eight Steps to Getting Design Right process to some project of your choosing. The project could be based on an idea of your own, or you may choose from some ideas we have created. In an appendix to this book you will find product design challenges for the following concepts:

- ◼ a bathroom-cleaning robot;
- ◼ a home health-care monitoring system;
- ◼ a night-vision system for automobiles; and
- ◼ an Internet-based meal delivery system.

No matter what project you work on, the goal of the exercise should be to create an initial product design and project plan. You should stop short of engineering-level detail. Even so, it is a substantial effort to carry out all the steps of the process. You should be prepared to devote many hours to the exercise.

Is It Worth the Effort?

This is not an easy book to read and these are not trivial concepts to master. Applying these techniques to a design problem requires effort and thought. Are this book and this approach worth all the effort to read, master, and apply? Absolutely! We have seen the benefits to students and ourselves from applying these ideas. Using a formal process to approach a design problem can seem awkward at first, but somewhere along the way you catch something that you would not have thought of otherwise, and, suddenly, you become a believer. On the second or third design problem, you find yourself using the techniques instinctively: They become the natural way to approach design.

What can you hope to see as a result of working your way through this book? If, presently, you do not consider yourself to be a designer, you may see your inhibitions to design disappear. You may discover a talent for design that you did not know you possessed. You may see your creativity unleashed. You may also find that your project leadership skills improve. Other members of your project team will start to look to you for leadership because you always seem to know what to do next, what pitfalls to avoid, and how to overcome obstacles. Even if you are an experienced designer, you may discover steps and techniques in this book of which you were unaware.

Why Use a Tabular Approach?

For the most part, the design approach we describe uses simple text-based tools and spreadsheet software (Microsoft Excel™). Recall that a system is a set of objects in relationship with another. A table is a standard way to represent relationships. Every cell of a table (a cell is the intersection of a row and a column) can be used to form a sentence relating the row heading to the column heading by means of the cell entry. For example, Table 1.1 is a simple description of a solar system showing the relation of moons to planets ("moons orbit planets") and planets to the sun ("planets

Table 1.1   Solar System in Tabular Form

	Sun	*Planets*	*Moons*
Sun		is larger than	
Planets	orbit		are larger than their
Moons		orbit	

orbit the sun"). It also captures relative sizes: "The sun is larger than any planet" and "the planets are larger than their moons."

Tables are our fundamental way to represent systems. The row and column headings will correspond to some subset of the objects of a system. The table entries will express the relationships. Because there are many different types of relationships, there will be many different tables.

Spreadsheet programs such as Microsoft Excel make it easy to create and manipulate textual lists, tables, and matrices (a *matrix* is a table of numbers or symbols). We make extensive use of Microsoft Excel in describing techniques for getting design right.

Excellent graphical techniques to describe systems are available. For example, Figure 1.4 displays a way to convey exactly the same information as that in Table 1.1 but in graphical form. This style of diagram is called an *entity–relationship diagram*. Objects ("entities") are drawn as nodes, and relationships are drawn as lines connecting the nodes. Lines are labeled with verb phrases, with the phrase positioned nearest the node that will form the subject of the sentence. Thus, "orbit" is placed near the node "planets" on the line connecting "planets" with the "sun" to suggest the relationship "planets orbit the sun."

Graphical representations have great visual power. It is likely that you prefer the representation of Figure 1.4 to that of Table 1.1. Why have we chosen the tabular approach over the graphical approach? The reason is that graphical presentations tend to be more time consuming to create and to make attractive than tabular views. You will find it requires a great deal of effort simply to apply the design steps of this book. It would be an extra burden to require you to complete the design steps and to present the results in attractive graphical form. We are content with having you present your work in a tabular format. We are therefore sacrificing visual power for ease of use.

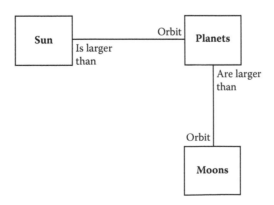

Figure 1.4   Solar system relationships in graphical form.

Graphical techniques are used extensively in advanced systems design, but there is a bewildering array of graphical styles to master (node shapes and meanings, line styles and meanings). We emphasize mastering the tabular technique first. As you discover different graphical presentations, you will be able to relate them to the underlying tabular views. Furthermore, advanced software packages for systems design can merge graphics, tables, and databases. Learning to use these sophisticated packages should be relatively easy after you have learned the simple tabular approach of this book.

The Getting Design Right Web Site

There are numerous spreadsheet files associated with this text. They can be downloaded by accessing the "Getting Design Right" Web site. The starting point is http://systemseng.cornell.edu/gettingdesignright/index.html. Follow the links there for the textbook, edition, and chapter in which you find the file referred.

Required Spreadsheet Skills

To do the exercises in this text, you will need to be familiar with Microsoft Excel (MS Excel™) or a spreadsheet program that can read MS Excel files. The following is a list of basic spreadsheet skills that will come in handy:

- copy transpose;
- drag and drop cells;
- merge cells;
- reorder rows and columns of a matrix;
- sum rows or columns;
- sort rows;
- create charts;
- use lookup formulas;
- compute sample mean and standard deviation; and
- display a Gantt chart.

If you are not familiar with these techniques, there is a tutorial, together with a fully worked example, available for download from the "Getting Design Right" Web site. The text will make reference to these skills prior to requiring them, so you can acquire the skills as you need them. Look for references in the text to a "spreadsheet skill."

To the Instructor: Where This Text Fits

This text was written to support a new type of course and fulfill a new need in engineering education. Traditional engineering curricula place a heavy emphasis on analysis and require a high level of mathematical maturity. Reclaiming the role of design and synthesis in engineering is an ongoing challenge. Design projects and design courses are enjoying a resurgence in engineering schools. But, in spite of the many project experiences that the modern engineering student now acquires, he or she usually is still expected to pick up the principles of design and project work on his or her own.

Having developed a graduate-level curriculum in systems engineering and having taught many industrial short courses on these topics, we recognized that the basic principles do not require years of preparatory mathematics and analysis. They can be taught in much the same way the arts are taught: by example. Why not make these principles available to engineering students at the beginning of their undergraduate studies? By extension, why limit the course to engineering students? Design is ubiquitous and there is nothing in this text that is inaccessible to the arts student, the hotel school student, or the agriculture major. On the contrary, we have seen students from other backgrounds excel in this course. We have adopted a style of instruction that works with mature audiences in professional settings as well as with freshmen with limited design and project experience.

Essentially, this text is an introduction to systems engineering, though we consciously resist using the term until the final chapter. It is a radical departure from systems engineering texts in that it delays the discussion of the complexity of systems until late in the book. The fundamental concepts of systems engineering find their way into the text, but the order in which they are introduced is new. We lay primacy on the basic design cycle and introduce the level-by-level decomposition of systems only in the final step of the cycle, "iterate the process." For the instructor who feels strongly that iteration pervades all steps of the cycle, it would be a simple matter to cover the last chapter first, at the beginning of the course, and thus permit a discussion of levels and iteration throughout the course.

What is missing from the text, compared with other texts on design, are domain-specific topics, such as design for manufacturability (mechanical engineering) or design for security (software engineering). Instead, we introduce the concept of design for secondary uses. This topic becomes a placeholder for the instructor to tailor the course to his or her particular expertise and interest.

Two chapters may come as a surprise to design instructors. The first "Develop the Architecture," represents a blending of different threads in the design and systems engineering literature. It does not figure strongly in the Design for Six Sigma (DFSS) literature. It comes instead from the systems and software engineering literature. Even here, however, our approach is different from others. We support the object-oriented methodology that has proven so successful in software engineering, but we approach the topic from a behavioral perspective. The object view emerges from the behavioral view.

The second chapter that may come as a surprise is "Execute the Design." Project management and project scheduling are typically removed from design and, instead, covered in courses on project management. Our intent is to present a more holistic view of design. This chapter pushes the limit on mathematical complexity in a general purpose text, but by treating resource-constrained scheduling as a form of computer puzzle (a downloadable spreadsheet), we have maintained the accessibility of the concepts.

The product and service design challenges included in the appendices were developed for our graduate engineering curriculum in systems engineering. We have also used some of these cases in our industrial short courses. There is nothing, however, that prevents their use with an undergraduate audience. They are useful design challenges no matter what the student's level of experience and preparation is.

The text can stand on its own as a basic course on design. It can serve as an introductory course on systems engineering, and it can serve as a supplementary text in a graduate-level systems engineering program.

Discussion

1. Debate the adage "the customer is always right" in the context of the *Vasa* example. At least one person should defend the king's position.

References

Fossberg, K., and H. Mooz. 1992. The relationship of systems engineering to the project cycle. *Engineering Management Journal* 4 (3): 36–43.

Love, T. 1993. *Object lessons: Lessons learned in object-oriented development projects.* New York: SIGS Books, Inc.

The *Vasa* Museum (brochure). The National Maritime Museum. Vasa Museet: Box 27131, 102 52 Stockholm.

Chapter 2

Define the Problem

The beginning is the most important part of the work.

Plato, fourth century BC

Introduction

Design projects can go awry at the beginning in at least two different ways. The first way is by rushing to design, that is, by jumping into designing a solution for a problem that is, in reality, poorly understood. Someone has expressed a problem ("Our customers complain that our software is confusing"). Someone with expertise has diagnosed the problem ("The customers haven't read the manual"). And someone generates a solution that logically solves the problem ("We should add a help button to every user screen of the application that will take them to the manual"). Solutions generated in such a manner may or may not solve the underlying problem. A lot depends on the hidden, internal thought processes of the participants and their success at communicating needs and ideas. More than likely, however, the rush to design will result in critical steps, such as gathering information about the problem, to be omitted. The opportunity to discover user-pleasing solutions will be lost. It is easy in large organizations for everyone to assume that these critical, up-front steps are someone else's responsibility. It is better if everyone takes at least part of the responsibility.

A second way in which design projects go awry is by a failure to start. Projects, as initially conceived, can be so vague that no one involved knows where to begin. A lot of time can be wasted while individuals wait for instructions on how to proceed. As pressure mounts on the design team to show results, the temptation to make assumptions and rush to design increases with it. Every project needs some basic questions answered before meaningful design work can proceed. The most basic question is "What are we trying to accomplish?"

This chapter is about getting started and laying the foundation for a successful design project. It is about defining the design project so that everyone involved has a clear picture of the purpose and scope of the project. It is about naming the problem, understanding the context of the problem, and looking at the problem from many different viewpoints. It is also about defining the character of the solution: what the solution must accomplish.

There are three major steps to defining the problem in the process of Getting Design Right:

1. Define the project.
2. Define the context.
3. Define the functional requirements.

We treat each of these steps in turn. We begin with defining the project.

Define the Project

There are five steps in defining the project:

1. Select the project.
2. Sketch the concept.
3. Define and tailor the process.
4. Identify the owner, the customer, and the user.
5. Write a mission statement.

Each of these steps is treated in a separate section.

Select the Project

The design project starts with an idea. It can be a problem you want to solve or it can be an idea for a solution to a problem. It can simply be an idea for creating something that appeals to you—that is, an idea for a "cool thing." If you start with a problem (a "top-down" approach), the process we use will help you to generate and explore ideas for a solution. If you start with a solution (a "bottom-up" approach), the process we use will require you to make clear what problem you are trying to solve and to consider other solutions. If it is a cool thing, you will be required to think through what makes it appealing. Therefore, do not be concerned about getting the initial idea right. What is important is to identify a problem or idea that you think has value. You will be spending a considerable effort in design, so you want to start with something that really motivates you.

From a practical perspective, if you truly intend to implement your design, there are considerations that should limit your choice of a project. You should be selecting something that is within your domain of expertise and that you have the time and resources to complete. But the history of invention and design is filled with examples of individuals who ignored this advice and, through perseverance, achieved success. Furthermore, in this book, we stop short of detailed design, so we are not constrained by engineering. We are illustrating a process that is appropriate no matter how ambitious your idea.

Name the Problem

Inventions often start with an "itch"—an irritation that something is not right with the world. Quite likely, others have felt the irritation as well, but most people just put up with it. In many cases, the irritation is preverbal: It does not have a name; it is just a feeling. The moment it becomes verbal is when the problem is named. Someone says, "I hate it when… or "I wish…" or "What we need is…" and puts his or her frustration into words. This is an important moment in the life of a design idea. Though no solution has been found, once a problem has been named, the possibilities for design begin.

Water Bottle Nipple Adapter

Consider the following story of the origin of a new product and note how the inventor went from itch, to naming the problem, to inventing a solution, to commercializing a product: "Inspiration struck Mr. Habeeb on a sweltering afternoon in Dallas, though, when he found himself with a cold bottle of water and no way for the baby slung across his chest to drink from it" (Rosenbloom 2007).

What Mr. Habeeb realized was that he wanted a way to attach a standard baby bottle nipple to commercially available water bottles. Once he had named the problem, it was a process of design and engineering to create a coupling adapter to fit water bottles and to which a standard baby bottle nipple could be attached. He then commercialized the invention. The result was the water bottle nipple adapter (Figure 2.1).

Solving a Noise Problem

Bruce Corson is an award-winning architect and consultant in collaborative problem solving based in California. He shares the following story about "naming the problem":

I was hired as a planning and design consultant by the directors of a large youth camp/environmental education center. They were embarking on an ambitious capital

Figure 2.1   Water bottle nipple adapter by BabySport.

improvement project and were at a point of crafting strategies for the project's approval and funding. It had been over 10 years since their last formal engagement with the public approval process, and, during this time, many of the neighboring parcels had changed ownership. Neighbors with years-long relationships with the camp had sold out to wealthy retirees. These newcomers had no knowledge of the camp—but they did have expectations of a tranquil life in the midst of beautiful natural surroundings.

The camp staff had been receiving an increasing number of calls over the previous few years from these new neighbors "complaining about the noise" created by the camp. The directors were now concerned about how these complaints might impact the approval of their capital projects.

They realized that their hoped-for approvals might well be jeopardized by unfavorable input from these neighbors. Hoping to avert this possibility, they had resigned themselves to undertaking an extensive proactive response to their "noise problem." They were already certain this would require:

■ the elimination or major modification of some of their more "energetic" programs: team-building activities, chanting, singing, sports, etc.;
■ the relocation and/or rescheduling of other activities; and
■ the building of major sound walls to screen activities not amenable to modification or relocation.

These changes—all assumed from the outset to be necessary—would be very expensive and would seriously compromise the camp's mission.

On arriving at my first meeting with the directors, I asked the questions:

■ Why have you named the problem "noise"?
■ Is that the best possible name for the problem?

To probe these first two questions, I asked a third question: What is "noise"?

My point was that the words we use to describe what we perceive can, in fact, constrain what we are able to perceive. In this case, by consciously reflecting on the meaning of the word "noise," we could probably find a different and much more elegant way to name the problem. (As the so-called "solution" can never be better than the so-called "problem" it purports to resolve, how we name the "problem" is the most critical act in any design process.)

Dictionary definitions of noise are

■ a sound that is loud, unpleasant, unexpected, or undesired; or
■ any disagreeable sound.

In physics, noise is a disturbance that reduces or obscures the clarity of a signal.

Initially, the camp directors were simply assuming their neighbors' "noise" complaints to mean "loud and, therefore, unpleasant and undesired sound." In their thinking, "noise" was a physical fact and meant literally "too much energy transmitted as pressure waves through the air." They were placing "noise" on a scale with physical dimensions ranging from "inaudible" to "deafening," and in doing so, they were limiting their design options to the standard responses of acoustic physics:

■ reduce the source energy (i.e., modify camp activities to reduce the excitement—the energy—of the children);

- increase the distance between source and sensor (i.e., relocate the children to more remote parts of the property); and
- absorb the energy radiating from the source (i.e., build massive sound walls between the children and the neighbors).

However, through helping us reflect on what we really meant by "noise," these definitions revealed another, much more promising dimension of "noise"—the human dimension of *perception*. The scale on this dimension ranges from "desired/agreeable" to "undesired/disagreeable." The problem statement now became: "How can we shift our neighbors' perception of our children's energy from "feared and undesirable" to "recognized and desirable"? This was no longer a problem for architects and engineers; it was a problem for community organizers and diplomats.

From this revelation, a whole new set of initiatives was designed in which camp representatives were sent to all the neighbors:

- to introduce the camp and its mission of environmental education and social growth;
- to describe specific camp activities and why those activities produced the sounds they did (in order to convert the perceived sound from "disagreeable noise" to "agreeable signal");
- to discuss how camp sessions and specific activities were scheduled (to keep the sound of these activities from being "unexpected");
- to issue invitations to camp open-houses;
- to engage the neighbors in mentoring programs for the youth; and
- to solicit new financial support from the neighbors for camp activities.

Through this process of "(re-)naming the problem," we avoided solutions that that would have limited the camp's activities and diverted scarce financial resources into unproductive assets such as sound barriers. Instead, we moved into strategies that would bring the community together and increase the resources available to the camp.

Bruce Corson

Sketch the Concept

We encourage you to start with a sketch of your idea. No matter how poor the artistic quality of the sketch, it is important to begin the process of visualizing the idea. Design engages both hemispheres of your brain: the left brain, with its power for verbalization and logical reasoning, as well as the right brain, with its power for visualization and intuitive reasoning (Edwards 1999). This text emphasizes a logical, step-by-step, process-oriented approach to design. That is, this is very much a "left-brained" text. But, there are points in the process where you simply must rely on the right brain to supply new ideas. The more you can express ideas visually, with sketches, the more opportunity you give your right brain to participate in the design process.

It is likely that you will also discover a greater excitement for your idea once you have a picture for it. Even a clumsy sketch and the act of creating it can generate that internal excitement. Some have theorized that this excitement comes from the right brain waking up and being encouraged to participate.

Every team-based design project, even a multimillion dollar one, should have a sketch or diagram that conveys the project idea visually on a single page. It is the most rapid and effective way to convey the basic idea to everyone involved in the project.

Example: Sketching the Toy Concept

For this text, we use a single example to illustrate and integrate the concepts. While the process is robust enough to handle something as complex as the design of a spacecraft, we would get lost quickly in the details of such a task. We need something simpler: a toy example. As a grandparent thinking about what to use as a toy example, my mind leaps immediately to toys for my grandchildren. What better way to illustrate a focus on delight than to picture a 4-year-old with a new toy? Our goal in the central example, therefore, is to design a toy to amuse a 4-year-old child and to delight everyone who comes into contact with it.

The first idea that comes to mind is that of building some sort of spring-loaded catapult. It just seems like a fun idea: We can picture getting down on the floor and playing with a child using this toy. To get a sketch of the idea, we combined different clip art images using a paint software program. Fortunately, we were able to find an image of a parent and child with blocks and another image of a historical catapult. Figure 2.2 is the result of that effort.

Figure 2.2 brings the idea to life. The artistic quality is low because the perspective on the catapult drawing does not match the perspective of the block castle or the other blocks. Also, shadowing and drawing styles are inconsistent between the two images. Nevertheless, the combined drawing instantly communicates a number of messages. First, this is a father and son activity, or grandfather and grandson activity, played on the floor. Second, there is a creative element of constructing a castle, but there is also a destructive aspect of knocking it down. Anyone who has played blocks with a child knows that both these aspects are present. With a young child, the parent's pleasure is in building the castle. The child's pleasure is in knocking it down. Over time, the child develops the creative building skill in imitation of the parent. The catapult introduces a new dimension to the game. An element of skill is required to position and aim the device.

Figure 2.2   A concept sketch for a child's toy catapult.

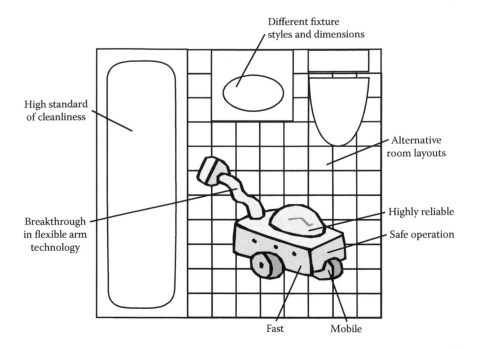

Figure 2.3   Annotated concept sketch of a bathroom-cleaning robot.

There is an unintentional consequence of combining the two images. The armed catapult in the drawing conveys a sense of threat, danger, and suspense. The drawing instantly raises a safety issue because the child is in the line of fire of the catapult. The issue of safety thus appears very early in the design process. We consider the safety issue formally as the design process proceeds.

Annotate the Product Sketch

There are many different styles of product concept sketches. We created the sketch in Figure 2.3 to describe a bathroom-cleaning robot (Appendix B). The *product concept sketch* is annotated with key features. The concept is for a safe, highly reliable, fast, mobile robot to exploit a breakthrough in flexible arm technology to clean fixtures with different styles and dimensions in alternative room layouts to a high standard of cleanliness. If we were successful in creating such a robot, these features would become the main selling points in our advertising campaign.

Define and Tailor the Process

How much time and effort do you want to spend up front on this process of getting design right? As you will discover, it is an open-ended process. There is no end to the questions you can ask and the analysis you can conduct. If you have a budget of time or money that limits your effort, you could find that you have spent it all on "getting it right" and have none left actually to "get it." You could spend all your effort planning what you are going to do and have nothing left with which actually to do it. That would be a foolish outcome. Therefore, decide up front how much of this process you will use and how detailed you intend to be. This is called *tailoring the process*.

The process, in its entirety, can be seen from the table of contents to this text. You may elect to skip some of the steps because they are not applicable to your design problem. Be aware, however, of the danger in skipping steps. This process has evolved over many decades of systems design activity by a very large community of professionals. It is a process that is generally applicable. We recommend that, unless a step is clearly inapplicable, you should attempt to apply it to your situation. Simply limit the effort you devote to a step based on the benefit you expect to receive. The full set of outputs from the process will be presented later (Table B.1 in Appendix B). The detail in these documents can be voluminous. Here, again, you should establish reasonable limits on the level of detail to which you commit yourself.

The best guide to tailoring the process is experience. As you use this process repeatedly, you will discover the effort required for performing it and the benefits derived from following it. You will also experience failures, and you will trace those failures back to the aspects of the process that you should have done better or spent more time on.

Example: Tailoring the Toy Example

Designing a toy is not difficult. We can amuse a child by drawing a face on a clothespin and showing her how to play with it. It may not be worth it to spend many hours in a conceptual design phase for a toy that can be built in a few hours and, furthermore, will be used by only one or two children. It is also not worth it to create numerous reports documenting the process and the design because, as is likely, they will be discarded once the toy is built. Nevertheless, we will apply all of the steps of the Getting Design Right process to this simple example. That is, we are deliberately overapplying the process relative to the benefits in order to show a fully worked example of the process. Thus, we are tailoring the process to meet our needs.

For the most part, we want to include the documentation as part of this text and have all diagrams and tables fit on a single page each. That is not always possible, but when tables have been omitted, they are still available as downloadable files on the Getting Design Right Web site.

Identify the Owner, the Customer, and the User

Who are you trying to please with your design? Sorting out the different requirements for a product or system can become confusing. It is good to start with a clear idea of who is involved and what their point of view is. Requirements for the system can then be traced back to a particular individual or point of view. Remember that our goal is to delight everyone who comes into contact with the design. Depending on the product, many different individuals could be involved. We will distinguish only three different roles at this stage: the owner, the customer, and the user. The *owner* of the system is the individual who sets design goals and authorizes major design decisions. The *customer* is the individual who approves purchase of the system and transfers it into use. The *user* is the individual who actually uses the system for an approved purpose. The same individual could play all of these roles, or multiple individuals could fulfill each role and whole agencies of people be involved.

It is helpful to think of the three roles of owner, customer, and user because they define useful viewpoints of the product or system. Looking at the product from different *points of view* is a systematic way to discover design requirements. For example, the user struggles with using the product to accomplish some task. This viewpoint gives rise to usability requirements. The customer examines the system from an economic viewpoint: Does the aesthetic or functional value of the system justify the cost? This viewpoint gives rise to economic requirements. Finally, the owner may have both economic and strategic goals in mind and may have special constraints on how the product can be built. This viewpoint can result in design constraints.

Example: Owner, Customer, and User of the Toy Catapult

For the toy catapult example, the primary user would be the child. However, as the concept sketch (Figure 2.2) illustrates, playing with the toy is pictured to be a father and son activity, so both parent and child can be visualized as users. The customer is difficult to identify in this example. As a grandfather, I imagine giving the toy as a gift. I am, therefore, the person making the economic judgment that the time, effort, and expense of creating and building the toy are worth the pleasure the child and parent may derive from it. In doing so, I am playing the role of customer. On the other hand, it is the parent who will accept the toy and decide if it is safe and appropriate for the child. That is, the parent transfers the product into use. Thus, the parent plays part of at least two roles: the customer and the user.

To simplify our thinking, let us think of the parent primarily as the customer and the child primarily as the user. That leaves the role of owner to be identified. This is clearly my role because the concept is mine, I will make nearly all of the design decisions, and I will impose important constraints on the design. For example, I have a limited set of tools in my workshop and I possess a limited set of hobbyist skills. Because I intend to design as well as build the toy, the design will have to be compatible with my skills and toolset.

Infant Crib Mobiles

Changing your point of view of a system can result in surprising insights. For example, it is apparent from studying old designs for infant crib mobiles that they were designed and marketed to appeal to the parents purchasing them. A parent's view of the mobile would look something like that shown in Figure 2.4.

The mobile is colorful and consists of shapes interesting to the eye. However, the view of the same mobile from the perspective of the infant lying in the crib underneath the mobile would be something like that in Figure 2.5.

The infant is looking up and seeing only the feet of the clowns: a much less interesting view! From this point of view, it is only the movement of the mobile that is interesting. The infant, who is the intended user of the product, has been neglected. Modern designs of mobiles correct this flaw by making the hanging objects more interesting from both perspectives, that of the parent and that of the child. The animals in the mobile in Figure 2.6 are angled so that both parent and child can enjoy them.

It is important, therefore, to consider multiple viewpoints in the design process—especially viewpoints of users who may not be present during the design discussions.

Other Categories of Individuals Affected by the System

The categories of owner, customer, and user are useful roles or viewpoints, but they are not exhaustive. For example, the role we have called "the owner" could be further categorized into roles of client, architect, and builder. The *client* is the individual who expresses the high-level goals of the system and

Figure 2.4   Traditional design for an infant mobile.

Figure 2.5   Infant view of traditional mobile.

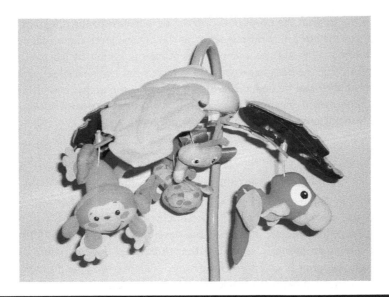

Figure 2.6   Peek-a-Boo Leaves Musical Mobile by Fisher-Price.

measures the success of the system. The *architect* translates the desires of the client into specifications for the builder to execute. The *builder* constructs the system according to the architect's specifications. The client's goals are likely to be vague and imprecise. The architect's specifications should be unambiguous. The builder may make low-level design choices, but they will be highly constrained by the architect's specifications. We have drawn the language of client, architect, and builder from the building and construction industry, but analogues to those roles exist in other industries as well.

In addition to users, we have *unintended users*—individuals who are affected by the system's existence or operation in ways that are not part of the primary function of the system. For example, if a highway is constructed near your home, you will hear the noise of traffic. You are affected by the system even if you do not own a vehicle. In this case, you are referred to as an unintended user. When you consider the full life cycle of a product or system, including its eventual disposal, you can identify many unintended users. Think, for example, of residents living near a landfill. Everything in the landfill was once part of a working, active, or useful product. These landfill neighbors are unintended users of all of these products.

The purpose of identifying all of these different roles is to give you the opportunity to look at the system from multiple perspectives. Every new role you identify gives you a different perspective and may give rise to new requirements on your design. The three primary roles we will consider in this text are owner, customer, and user. They constitute a minimal set of perspectives.

Write a Mission Statement

The *mission statement* is a succinct statement of the purpose or goals of the system to be designed from the perspective of the owner or the next higher agent in the design management hierarchy.

Our perspective in systems design is typically that of the designer or architect. Because of the iterative nature of systems design, larger systems are decomposed into smaller systems, and the process continues until we finally reach the definition of subsystems that can actually be built. At each level in this iterative process, some designer is acting as the representative of the next higher agent in the design chain to ensure that the needs of the owner are reflected in the design of the subsystems. That next higher agent may be another designer or, at the highest level, the actual owner.

Because the designer is really the agent representing the owner in the design management process, it is important for the designer to be able to state what the purpose or goals of the owner are, to the extent that they can be deduced at this level in the hierarchy. It is also useful if you, as the designer, state the mission of each subsystem under your design control so that designers at the next level have a clear understanding of their design purpose.

Boeing 777 Mission Statement

The Boeing 777 is a wide-body jet, midsized between the 767 and 747 jets, with two engines, and it is capable of transoceanic flights with higher fuel efficiency than its predecessors and competitors. The 777 was launched in 1995 and quickly became one of Boeing's best-selling aircraft. Its design and development in the late 1980s and early 1990s came during an industry downturn that threatened the company's future. The 777 was a high-risk project, and it was following a major development project, the 7J7, that had been cancelled because of lack of customer interest.

In designing the 777, Boeing was determined not to repeat the mistakes of the 7J7 and committed itself to working together with its customers to design an aircraft that the customers truly wanted. The company started by creating a customer focus group consisting of representatives from eight of its major customers. The so-called "Gang of Eight" met with Boeing designers three times in 1989 and 1990 to establish general design goals.

In 1990, Boeing won a contract from United Airlines to build the first 34 aircraft in the 777 model line. This was an order for an aircraft that had not even entered the detailed engineering phase. It won out against aircraft from McDonnell–Douglas and Airbus that were either already flying or at least in final test. United was taking part in a major gamble.

Within hours of United's decision to purchase 34 firm and an additional 34 optional aircraft, a pivotal event occurred that set in motion a phenomenal cultural change within Boeing and United that dramatically changed the tone and texture of Boeing's future relationships with its customers (McKinzie 1996).

Jim Guyette was United's executive vice president of operations in 1990. He had chaired the marathon 70-hour sales session in which Boeing emerged as the contract winner. He realized the magnitude of the gamble that United was taking and wanted to be convinced that Boeing would live up to its new-found commitment of "working together" with its customers. United's recent experience with design and production delays on the Boeing 747-400, a sophisticated upgrade to the 747 model, had been painful. Guyette took out a pad of yellow legal-sized paper and quickly drafted a list of objectives for the 777 project. He then took the one-page handwritten document to the Boeing executives for their signatures. Signing the document for Boeing were the head of the New Airplane Program, Phil Condit, and executive vice president, Dick Albrecht. It was a dramatic moment as these two executives committed their company to a new form of partnership with its customers. The subsequent success of the 777 can be traced to this mutual commitment of United and Boeing. This is the mission statement they agreed to:

In order to launch on-time a truly great airplane, we have the responsibility to work together to design, produce, and introduce an airplane that exceeds the expectations of flight crews, cabin crews, and maintenance and support teams and ultimately our passengers and shippers. From day one:

- best dispatch reliability in history
- greatest customer appeal in the industry
- user-friendly and everything works.

Condit, Guyette, and Albrecht 1990. Reprinted with permission from the Boeing Company

Example: Mission Statement for a Toy Catapult

Our mission in this example is to design a toy catapult to amuse a 4-year-old child safely, creating an opportunity for parent and child to play together in a way that delights them both.

Note that we have committed ourselves to the design of a particular type of toy: a toy catapult. We could have stated the mission more broadly in terms of the desire to design a toy. That would have resulted in a larger design space to explore. Our purpose in selecting the particular example of a toy catapult is to keep the design space small for illustrative purposes. We also selected this type of toy because it exhibits a non-trivial mechanical function that requires several subsystems to work together. It can thus be used to illustrate system decomposition.

Define the Context

There are five steps in defining the design context:

1. Define the system boundary.
2. Document the context of the system.
3. Study the current context.
4. Collect customer comments.
5. Summarize the project (product) objectives.

Define the System Boundary

Design projects frequently run into trouble because different individuals on the project conceive of the system in very different ways. If there is only one person engaged in the design, then communication is not an issue. However, as soon as a team of people is engaged, there is potential for miscommunication. This potential difficulty increases geometrically with the number of people needing to communicate with each other. For team projects, therefore, it is important to state and clarify continually what the *system boundary* is—that is, what is within the scope of the design and what is outside the design scope.

Entities that are external to the system have their own goals, functions, behavior patterns, and interfaces. *External entities* interact with the system but are not subject to design change. The system or product under design must relate to these external entities on their own terms. Confusion over what entities are external can be avoided by clearly identifying them from the beginning.

Whenever a new entity is identified, it should be classified quickly as internal, external, or irrelevant. If an entity is not related or only weakly related to the system, it is irrelevant and can be ignored.

Example: External Entities of the Toy Catapult

The focus of the toy example is on the toy itself. The toy might include labels for its use and initial packaging but nothing else. For example, we are not planning to design a storage container for the toy. Furthermore, at this stage of conception, we do not think of the projectiles that the catapult will throw as being part of the system. That could change. At a later stage of design, we could decide to include custom-designed projectiles. For now, however, they are external entities. Similarly, the parent and child are external to the "catapult system." Thus, we have identified parent, child, projectiles, and storage container as external; and toy, labels, and packaging as internal.

What about entities that are not related to the toy? For example, the child's car seat has no relationship to the toy. It is irrelevant; it is not mentioned.

Imagine the confusion and wasted effort if this were a team project and different individuals made different assumptions about where the system boundary lay. Suppose the toy designers changed the toy dimensions without realizing that someone else was designing a storage container and the storage designers needed to know the new dimensions. We could end up with a storage container that did not fit the toy. It is better to be clear up front as to whether or not the system includes a storage container. In our case, we have excluded the storage container from the system.

Document the Context of the System

A *system* is a collection of objects in relationship to one another. Having identified our concept system and the entities that are external to it, we should take the time to capture the relationships between our system and these external entities. That is, we want a view of our proposed system in the context of a larger system. The purpose is to explain why each external entity was thought to be relevant to the system. At this stage, we are making note of relationships that we should explore in greater detail later. Associated with each relationship is an interface that will need to be specified. In this text, we will use both context diagrams and context matrices to document the context of a system.

Context Diagrams

A *context diagram* is a concise visual network of the relationships between entities in a system. The nodes of the network are the proposed system and the external entities. The connecting arcs of the network are the relationships between the system and the external entities. The diagram may include arcs between external entities, but because these are outside the design scope they serve only as background information.

The context diagram is drawn as a network of rectangular nodes and rectilinear (right-angled corners) arcs. The nodes are labeled with the names of system elements (nouns). The arcs are labeled with relationships (verb phrases). It should be possible to form a simple sentence using the label on an arc. By combining the noun of the tail node as subject, the verb phrase on the arc, and the noun of the head node as object, the resulting sentence should describe a relationship of the

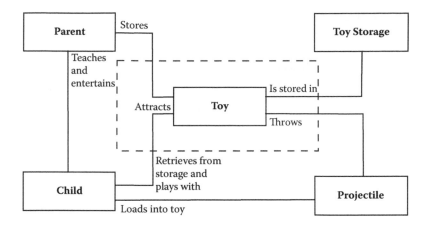

Figure 2.7   Context diagram for the toy catapult.

tail to the head. The arcs in a context diagram do not use arrowheads. Instead, the direction of the relationship is indicated by placing the label close to the tail node (the origin node).

Example: Context Diagram for the Toy Catapult

Figure 2.7 is a context diagram for the toy catapult. Observe the format of rectangular nodes and rectilinear arcs. The arcs do not have arrows. Instead, the verb phrases on an arc are located closest to the node containing the subject of the relationship. For example, "is stored in" is located close to "toy" to form the sentence "Toy is stored in toy storage." If it were located close to "toy storage," the resulting sentence would be "Toy storage is stored in toy," which does not make sense. Also, note that the same arc can be used for two different relationships, as in "Toy attracts child" and "Child retrieves [toy] from storage and plays with toy."

Some relationships are symmetrical: "Toy storage stores toy" is symmetrical to "Toy is stored in toy storage." Only one such relationship needs to be recorded; the other is implied.

The complete set of relationships captured by Figure 2.7 is as follows: "The parent teaches and entertains the child. The parent stores the toy. The toy is stored in toy storage. The toy attracts the child. The child retrieves the toy from storage and plays with it. The child loads a projectile into the toy. The toy throws the projectile." The dotted line in Figure 2.7 is used to mark the system clearly. Everything outside the dotted line is an external entity.

Context Matrices

A *context matrix* is a concise tabular presentation of the relationships between entities in a system. The rows of the table are the entities of the system, listed in any order. The columns of the table are the entities of the system listed in the same order as the rows. The entries of the table are verb phrases connecting the row entity as subject of a relationship with the column entity as object of the relationship. A context matrix conveys exactly the same information as a context diagram.

Table 2.1 Context Matrix for Toy Catapult

	Parent	*Child*	*Toy*	*Projectile*	*Toy Storage*
Parent		teaches and entertains	stores		
Child			retrieves from storage	loads into toy	
Toy		attracts		throws	is stored in
Projectile					
Toy storage					

It does so simply in a tabular rather than a graphical form. No special formatting is required for a context matrix apart from the requirement that all text should be fully visible (word wrap should be enabled for each cell in the matrix). A context matrix is easier to create than a context diagram but it is not as visually appealing.

Example: Context Matrix for the Toy Catapult

Table 2.1 displays a context matrix for the toy catapult example. It contains exactly the same information as the context diagram in Figure 2.7. For example, we read that "the parent teaches and entertains the child." The row entity is the subject of the relationship. The column entity is the object of the relationship. Observe that there are no relationships in which the projectile or toy storage is subject (these rows are blank).

The border around the cell at the center of the matrix (toy row and toy column) plays the same role as the dotted line in Figure 2.7: It marks the system boundary.

Study the Current Context

Contextual inquiry (Beyer and Holtzblatt 1998) refers to the activity of gathering data on the context in which a system is going to be used. It is typically accomplished by visiting customer work sites, observing the current process, and collecting data on customer needs and wants.

When you conduct a contextual inquiry, it is important not to present yourself as the "expert" or the "solution supplier." You are in neither an expert nor a sales role. Instead, your role is some combination of the following roles.

Naturalist, Anthropologist, or Observer

The *anthropologist* attempts to observe societies without changing them. So, too, the naturalist studies the habitats and behaviors of different animal and plant species. These scientists write in the third person ("They…") and document the way things are. Similarly, there is a role within contextual inquiry for passive observation and documentation: capturing the "as is" function of the system.

Like an anthropologist, the development team conducting a contextual inquiry wants to know not only the behavior but also the organization, culture, and values of the workplace under study. How are decisions made? Who makes them? What information is used? How is information communicated? What are the beliefs of the users? How does the culture sustain itself? What is important to the users? What pleases or displeases them? What is the location and environment of the work place? How fast are activities performed?

Apprentice, Questioner, or Interpreter

A more active role within contextual inquiry is to engage the user as someone who wants to learn the process or acquire the skill. The interviewer asks questions the way an *apprentice* would ("Why do you do that?" "How would I accomplish this task?"). The interviewer will also attempt to interpret events ("Does this mean…") or articulate general principles ("So, is it generally the case that…") and ask the user to correct or refine the ideas.

Children are often most natural in this role, feeling free to ask an endless series of "why" questions until the adult answers in exasperation, "Just because." As an interviewer, it is good to pursue a series of "why" questions to the fifth level ("the five whys") because it will typically take you to the deepest level of understanding a user has about any system.

Partner or Suggester

At some point, the interviewer may slip into less of an apprentice role and more of the role of a *partner*. The interviewer would then explore alternative ways to perform a task or achieve a result ("What if we did this?") and ask for the user's reaction. In this way, the interviewer will discover important constraints ("We can't do that because…") or the interviewer may discover a user-pleasing approach to a problem.

Pitfall of Contextual Inquiry

The pitfall of contextual inquiry is its potential to collect irrelevant data. Imagine, for example, that you were interviewing parents about how their child plays with toys. It would be easy to go off-topic and talk about many aspects of child-raising and child development: sleeping patterns, eating patterns, emotional needs, discipline problems, and so on. You could amass a large amount of information that is irrelevant to the mission of toy design. Much of that material would be discarded once you finally get to the task of interpreting the results of the contextual inquiry.

To guard against this tendency to be sidetracked, the interviewer needs to begin the process with an understanding of the mission and an identification of the user interfaces that need to be explored. A good context diagram should be mapped out in preparation for this step. The information gathered must be relevant to the mission and the interfaces. For example, the context diagram for the toy catapult identifies a partial set of relationships to explore in an interview with parents (what appeals to the child, what appeals to the parent, how the parent plays with the child, how the child plays alone, how the toy is stored, etc.). Because the interviews would consider more than just toy catapults, the diagram should be extended to consider other toys.

Apart from the need to stay relevant, the focus of contextual inquiry should be kept on customers' activities, needs, and wants. As you will see, we will extract from these interviews the attributes of product design that are most important to the customer. Another use of contextual inquiry that is relevant for commercial design is to understand how purchase decisions are made and how group relations (e.g., office politics or marital roles) influence the purchase and use of certain products. For this text, we downplay this use and focus on the relation of the customer to the product.

A Little Slip of Paper

A company in the building products industry contracted with me to develop software to handle their vehicle routing and scheduling needs. I spent a day sitting beside the scheduler following him through the activities of his day. I played the role of apprentice, asking him to teach me how he did his job. For the most part, I anticipated correctly all the various tasks he performed based on my knowledge of scheduling (selecting orders based on geography, choosing a vehicle type, organizing orders into truck loads, and sequencing deliveries). However, the scheduler surprised me at one point by making a phone call and writing some numbers down on a slip of paper. I asked him what the slip of paper represented. He explained that he had just called the production manager to find out what the production rates on the various production lines would be for the coming week. I was surprised to discover that one aspect of the vehicle-scheduling job was to coordinate shipping volumes with production volumes for each production line. I would have missed this aspect had I not conducted a contextual inquiry. Having made the discovery, I integrated the production coordination aspect of the job into the scheduling software. This led to an example of user delight. The scheduler subsequently said that my software was great. "Everything I need," he said, "is right there on the screen."

Collect Customer Comments

There are different ways to collect customer data. One of the best ways, as described in the "Contextual Inquiry" section, is to spend time with the customer at the customer work site or, as Harley-Davidson does (see the Harley-Davidson case study in Appendix A), at rallies or other events where your customers gather. This form of interaction does not require that there be an existing product upon which you are gathering feedback. You could be seeking simply to understand the context in which particular tasks are performed. Out of this understanding, you will propose how a new system might meet the needs that you identify.

Imagine that your team has visited four to six different customer work sites and interviewed 15–20 individuals at each site who are engaged in different aspects of the function you hope to develop a system to address. Suppose that your interviews yield between 50 and 100 comments per individual interviewed. As you can see, this data collection process can very quickly yield several thousands of customer comments.

There are other techniques to collect customer comments. You can ask your potential users and customers to complete surveys. You can collect error logs, where users record problems they experience with an existing piece of equipment or software. You can comb through warranty claims on existing products looking for the types of ways in which products fail and what customers write about the failure. You can distribute prototypes of your product to potential users for free, in return for detailed product evaluations.

When you collect customer comments, try to get comments written or spoken by the users, using their words as much as possible. Surveys can be misleading because the survey writers compose the questions and restrict the possible answers. For example, when soliciting student feedback on my courses, I do not like to use multiple-choice surveys. I prefer to ask students to identify three things they liked about the course and three ways in which the course could be improved. Aspects of the course that students like or dislike frequently surprise me. If I had composed multiple-choice questions, I would have missed some of these responses. In particular, if students are upset about some aspect of the course, this will usually surface, in their own words, in response to the second question.

Example: Parental Comments about Toys

Table 2.2 is a collection of comments from parents about toys. None of the comments relate directly to a toy catapult but they are useful for learning what is important to parents. As you read the comments, note that they can seem to be repetitive and, sometimes, contradictory. That is not surprising because they come from different parents with similar concerns and differing opinions. At this point, they are simply unorganized comments. In the next section, we will organize them to summarize the "voice of the customer."

Summarize Project (Product) Objectives

Customer comments provide valuable, detailed insight into what customers value. However, in their raw form, they are too fragmentary to provide guidance to designers. After collecting these detailed comments, the next step is to summarize them into statements about design objectives from the user or customer perspective.

This process of summarizing customer comments is our first example of diving and surfacing. *Diving and surfacing* refers to the development of a detailed view of some aspect of a system ("diving") followed by summarizing the details into a more abstract view of the system ("surfacing"). In this case, collecting the customer comments is an act of diving. Summarizing them is an act of surfacing. We will continue this pattern of diving and surfacing throughout the design process. This pattern ensures that our abstractions are based on solid details, rather than on airy generalizations, and it keeps us from getting lost in those details.

There is a straightforward way to summarize a list of detailed comments. The *affinity process* is a two-step process to organize lists into named categories. The first step is to place the items on the list into groups based on affinities that the items seem to share. The second step is to write a name or sentence for each group that describes the affinity that characterizes the group.

The affinity process requires work. It is a process of data collection, data grouping, and data summarization. Unlike statistical data analysis, however, the affinity process deals in customer comments expressed in human language, and it encourages highly subjective grouping techniques. Mechanically, this can be done in different ways. We describe two techniques in the following sections: sticky notes and drag and drop.

Technique: Sticky Notes (Large Group Affinity Process)

In a full-scale design effort, the collected comments from customers could number in the thousands. Beyer and Holtzblatt (1998) suggest the following process. Assemble a team of designers to review the customer comments. Plan that this activity will take 1 day to assimilate the comments thoroughly. You should budget about one person per 100 comments for this activity.

Prior to the meeting, the individual customer comments should be written or printed onto sticky notes (adhesive-backed note paper) and posted onto the surface of a large wall (or around the walls of a conference room).

Table 2.2 Parental Comments for Toy Catapult

Parental Comments
The problem with grandparents is that they love to buy the kids toys with lots of little pieces. Guess who has to pick up all the pieces? It's not the grandparents.
Once the lever or switch breaks, the toy is useless. Usually that is the flimsiest part of the toy.
The battery-operated toys never get played with once the batteries run out. We never think to buy more batteries.
The noise of that toy drives me crazy.
My son loves toys with wheels. He spends ages moving them all around and lining them up.
The kids love it when I get down on the floor and play with them.
My daughter tells a story when she plays. She talks to the toys. They all have names.
We're running out of space to store all the toys. Some of them are huge.
I don't buy toys with sharp edges.
Sometimes, I will get a toy to do something silly. My son will laugh uproariously and then ask me to do it again. Then, every time I do it he will laugh and ask me to do it again. This goes on and on until I have to call it quits.
When some toys break, they shatter. The sharp pieces can be quite dangerous.
The baby is attracted to any toy with eyes painted on it.
I like giving them wooden toys with a natural finish. The texture and color are more soothing than plastic toys.
Some toys are way too complicated. The kids just push the buttons randomly.
Kids are excited by bright primary colors.
Transporting things seems to be fun. She will fill the wagon with toys and pull it all around the house.
If he sees someone else playing with his toy, he is more likely to be interested in it.
It is hard to teach children to pick up all their toys. At the end of the day, it is often me who goes around and tidies up the toys. Some toys are easier to put away than others.
You can keep him entertained by giving him a bucket and things to put into it.
A lot of toys with latches and doors are broken. Kids are pretty rough on that sort of thing.
They like repetitive games, like dropping a ball down a chute.
It's interesting how a child will not play with a toy for months and then suddenly it will become the favorite toy again.
A lot of the toys I find boring. I can't think of anything to do with them. It's no wonder the kids don't play with them.

Table 2.2   Parental Comments for Toy Catapult (Continued)

Parental Comments
If she sees me working with something, then she will run and find a toy that looks like it and imitate my actions. She has a toy calculator that she uses as a cell phone.
If it makes a noise, they like it better.
Most of the toys nowadays that do things are battery powered. They are great until the batteries run out and then they are junk.
The most interesting toys are the ones that do things. But they are usually broken.
As long as it has wheels, he will play with it.
I like toys that hold other toys. It makes them easier to stack and store.
If the toy has a name and a personality (e.g., Thomas the Train™), they are attracted to it. It's like playing with a friend.
It's funny how you give a kid a fancy toy and what he really wants to play with is the box it came in.
If I tell a story about the toy as I play with it, then they are more likely to repeat my actions.
To me, the feel of a toy is important. I like to find toys that have an interesting shape or texture.
I am usually glad when the batteries run out. The noise of these toys can be really annoying.
Once you lose the pieces to a toy, it is very hard to play with. Lots of toys are made so that only one piece fits (like the driver of a truck). Usually you can find the toy or its pieces, but not both.
If a toy does something interesting but is too complicated, my child will bring the toy to me to play with. We can have fun together with it, if I have the time.
I worry about the kids getting their fingers pinched in some of these toys.

The team of designers then spends time walking along the wall reading customer notes and rearranging them according to affinities that they detect. It is important that the designers not begin with a pre-established list of grouping categories. One of the purposes of the activity is to see if new categories emerge.

Designers are encouraged to put notes together "if it feels right," even if they cannot immediately articulate the relationship. Designers can also make copies of notes by hand if they feel a note belongs in multiple places.

As themes emerge, whole groups of notes should be moved together and arranged to express a hierarchy of customer concerns.

The design team can construct the hierarchy of notes in any way that serves its needs, but Figure 2.8 suggests a good model for rapidly conveying the sense of organization that emerges from the activity.

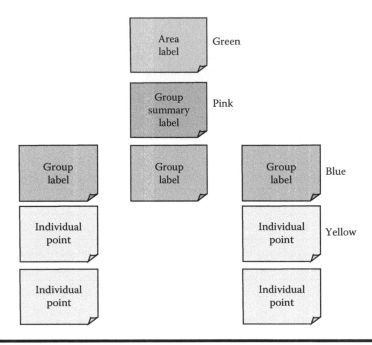

Figure 2.8 The affinity diagram. (Adapted from Beyer, H., and K. Holtzblatt. 1998. *Contextual design: Designing customer-centered systems*. San Francisco: Morgan Kaufmann Publishers.)

The Affinity Process Using MS Excel

For small numbers of customer comments (say 50 or less), it is effective simply to use MS Excel to organize them into categories. The drag-and-drop feature of MS Excel makes this possible (spreadsheet skill: drag and drop cells).

Table 2.3 illustrates three possible affinity groups that could be formed from the customer comments from the toy example. Figure 2.9 is a zoomed-out view of a complete solution for all 37 customer comments. We organized the customer comments into 11 affinity groups. The details of the groupings cannot be seen in this figure, but they can be found in the spreadsheet on the Getting Design Right Web site for this chapter.

Another technique is to merge cells to create hierarchical column headings (spreadsheet skill: merge cells). Table 2.4 shows a partial view of a hierarchy created for the customer comments.

Finally, we transpose and reformat the summary. Table 2.5 displays the complete summary of the customer comments. It is the *voice of the customer,* summarized.

Results of the Affinity Process

The principal output of the affinity process should be a list of categories and themes—expressed in customers' language—that identify their concerns, both major and minor. This can be published as a document (with the detailed customer comments) to be used as a reference throughout the design cycle.

A secondary but intangible output of the affinity process is a sense of ownership of the customer perspective by the designers themselves. They have spent a day putting themselves into

Table 2.3   Three Possible Affinity Groups

Group 1	Group 2	Group 3
The problem with grandparents is that they love to buy the kids toys with lots of little pieces. Guess who has to pick up all the pieces? It's not the grandparents.	Once the lever or switch breaks, the toy is useless. Usually that is the flimsiest part of the toy.	The battery-operated toys never get played with once the batteries run out. We never think to buy more batteries.
It is hard to teach children to pick up all their toys. At the end of the day, it is often me who goes around and tidies up the toys. Some toys are easier to put away than others.	A lot of toys with latches and doors are broken. Kids are pretty rough on that sort of thing.	Most of the toys nowadays that do things are battery powered. They are great until the batteries run out and then they are junk.
Once you lose the pieces to a toy, it is very hard to play with. Lots of toys are made so that only one piece fits (like the driver of a truck). Usually you can find the toy or its pieces, but not both.	The most interesting toys are the ones that do things. But they are usually broken.	

the shoes of the customer and paraphrasing what the customer is trying to say. They are likely to identify quite strongly with the perspectives they have heard.

The paraphrased customer concerns that are identified through this process can be validated through further interactions with the customer using survey techniques (e.g., "Would you agree with the statement that...").

Define Functional Requirements

Our goal now is to write a set of statements that list the *functional requirements* of the system—that is, the functions the system must perform or, in other words, what the system or product is expected to do. Functional requirements should be stated at an abstract or general level. They should specify what the system must do but not how it is expected to do it. The question of what to do is a high-level design decision; the question of how to do it is a lower level decision.

Requirements are written formally as "shall" statements. For example, "the system shall present the user with a menu of choices." Avoid using tentative language such as "should" because saying the system should do something is more of an objective than a requirement. Requirements are reserved to express what a system must do. If a design fails to satisfy a requirement, then it is, by definition, an unacceptable design.

Care must be taken in writing requirements because we do not want to eliminate what might be a potentially user-pleasing design as a result of being too restrictive in our requirements. On the other hand, we do not want to miss stating a requirement that could be crucially important to the

The problem with grandparent is that they love to buy the kids toys with lots of little pieces? Guess who has to pick up all the pieces? it's not the grandparents.

It is hard to teach children to pick up all their toys. At the end of the day it is often me who goes around and tidies up the toys. Some toys are easier to put away than others.

Once you lose the pieces to a toy, it is very hard to play with. Lots of toys are made so that only one piece fits (like the driver of a truck). Usually you can find the toy or its pieces, but not both.

Once the lever or switch breaks, the toy is useless. Usually that is the flimsiest part of the toy.

A lot of toys with latches and doors are broken. Kids are pretty rough on that sort of thing.

The most interesting toys are the ones that do things. But they are usually broken.

The battery-operated toys never get played with once the batteries run out. We never think to buy more batteries.

Most of the toys nowadays that do things are battery-powered. They are great until the batteries run out and then they are junk.

The noise of that toy drives me crazy.

I am usually glad when the batteries run out. The noise of these toys can be really annoying.

My son loves toys with wheels. He spends ages moving them all around and lining them up.

Transporting things seems to be fun. She will fill the wagon with toys and pull it all around the house.

You can keep him entertained by giving him a bucket and things to put into it.

They like repetitive games, like dropping a ball down a chute.

It's interesting how a child will not play with a toy for months and then suddenly it will become the favorite toy again.

A lot of the toys I find boring. I can't think of anything to do with them. Its no wonder the kids don't play with them.

As long as it has wheels, he will play with it.

Its funny how you give a kid a fancy toy and what he really wants to play with is the box it came in.

If it makes a noise, they like it better.

The kids love it when I get down on the floor and play with them.

If he sees someone else playing with his toy, he is more likely to be interested in it.

If she sees me working with something, then she will run and find a toy that looks like it and imitate my actions. She has a toy calculator that she uses as a cell phone.

If I tell a story about the toy as I play with it, then they are more likely to repeat my actions.

Sometimes, I will get a toy to do something silly. My son will laugh uproariously and then ask me to do it again. Then, every time I do it he will laugh and ask me to do it again. This goes on and on and on until I have to call it quits.

My daughter tells a story when she plays. She talks to the toys. They all have names.

The baby is attracted to any toy with eyes painted on it.

If the toy has a name and a personality (eg. Thomas the Train), they are attracted to it. Its like playing with a friend.

We're running out of space to store all the toys. Some of them are huge.

I like toys that hold other toys. It makes them easier to stack and store.

I don't buy toys with sharp edges.

When some toys break, they shatter. The sharp pieces can be quite dangerous.

I worry about the kids getting their fingers pinched in some of these toys.

I like giving them wooden toys with a natural finish. The texture and color are more soothing than plastic toys.

Kids are excited by bright primary colors.

To me, the feel of a toy is important. I like to find toys that have an interesting shape or texture.

Some toys are way too complicated. The kids just push the buttons randomly.

If a toy does something interesting but is too complicated, the child will bring the toy to me to play with. We can have fun together with it, if I have the time.

Figure 2.9 Eleven possible affinity groups.

Table 2.4 A Hierarchical Group of Affinities

Make the Toy Appealing for both Parent and Child			*Make the Toy Safe*
Make a toy with personality	**Make a toy pleasing to the eye and to the touch**	**Do not make a toy that makes an annoying noise**	**Do not let the toy injure my child**
My daughter tells a story when she plays. She talks to the toys. They all have names.	I like giving them wooden toys with a natural finish. The texture and color are more soothing than plastic toys.	The noise of that toy drives me crazy.	I don't buy toys with sharp edges.

design. The classic example of a missing requirement is a man who builds a boat in his basement workshop and then discovers that it is too big to carry up the stairs and get out through a door. One way of expressing the missing requirement in this case is "the boat shall be small enough to be carried upstairs and outdoors."

It can be difficult to write high-level requirements if you have not had any experience in the process. We use a "dive-and-surface" approach. That is, we encourage you to begin at whatever level of detail you are comfortable with and then dive into writing detailed requirements. Once you have a rich collection of detailed requirements, you will surface and write abstract requirements that summarize these detailed requirements. We find that abstraction is easier if you have a good collection of details from which to base your generalizations. The process you will use to form these abstractions is the affinity process: the same process used to summarize customer comments.

Table 2.5 The Voice of the Customer

Make the toy fun for both parent and child	Make a toy that interests my child
	Make a toy that involves me, the parent
	Do not make the toy too complex
Make the toy appealing for both parent and child	Make a toy with personality
	Make a toy that is pleasing to the eye and to the touch
	Do not make a toy with annoying noise
Make the toy safe	Do not let the toy injure my child
Make the toy easy to put away	Make the toy easy to put away (and find again)
	Make a toy that is easy to store
Make the toy playable for a long time	Make the toy reliable
	Avoid making a battery-powered toy

The primary output of this dive-and-surface approach is the collection of summarized, high-level requirements. You may think the effort spent on detail is wasted, but that is not the case. The detailed requirements are not lost in this approach. They can be reused later when we get to detailed analysis.

Another feature of this approach to discovering and defining requirements is that we focus on required behaviors of the system. A system *behavior* is a description of a sequence of events or activities from a beginning event to one or more possible terminal events. The beginning event is called the *stimulus*. Any one of the terminal events is called a *response*. The stimulus typically comes from an external entity (but, in more complex systems, can be internally generated). Some of the activities in the behavior may be performed by the system; others may be performed by external entities in response to actions taken by the system. By thinking through behaviors, we naturally discover functions that the system must perform. In particular, whenever we discover an activity in the behavior that the system performs, we write it formally as a functional requirement.

Our approach to defining the functional requirements for the design consists of six steps:

1. Collect use cases.
2. Prioritize use cases.
3. Describe use case behaviors.
4. Summarize functional requirements from use case behaviors.
5. Repeat for secondary use cases.
6. Finalize requirements.

Example: A Toy Catapult

The toy catapult example is integrated into the text of the next few sections because it is easier to explain the concepts by example. We begin the process with a simple context matrix, as shown in Table 2.6. This version of the context matrix is even simpler than the one shown in Table 2.1. We will progressively enhance the context matrix with new system entities as we discover them. The context matrix can get quite large as a result. Later, during the summarization phase, we will eliminate some of the detail and contract the size of the matrix.

Collect Use Cases

The first step in this dive-and-surface approach to defining functional requirements is to list a set of use cases. A *use case* is a situation in which a user would want to use the system. Usually, at least one use case is obvious, but more use cases come to mind as you think about the system

Table 2.6   Initial Context Matrix for Toy Catapult

Is Related to	Child	Parent	Toy
Child			retrieves and plays with
Parent	teaches and entertains		stores
Toy	attracts		

Note: The system boundary is indicated by the bold border.

Table 2.7 Primary Use Cases for Toy Catapult

Use Cases
Child chooses toy
Child retrieves toy
Child plays with toy
Parent teaches child
Parent puts toy away
Parent entertains child

from different users' perspectives. Remember the infant mobile example: The infant mobile can be viewed from the parent's perspective or the infant's perspective. In a team setting, this is an opportunity for brainstorming: Think of as many different use cases as you can. Do not criticize anyone's suggestion because criticism will discourage the free flow of ideas. Table 2.7 is a basic set of use cases for the toy catapult. It was derived from the initial context matrix in Table 2.6.

Patriot Missile Defense Failure

On February 25, 1991, a Patriot missile defense system operating at Dhahran, Saudi Arabia, during Operation Desert Storm failed to track and intercept an incoming Scud missile. This Scud subsequently hit an Army barracks, killing 28 Americans.

The Patriot is an Army surface-to-air, mobile air defense missile system (Figure 2.10). Since the mid-1960s, the system has evolved to defend against aircraft and cruise missiles, and more recently against short-range ballistic missiles. The Patriot system was originally designed to operate in Europe against Soviet medium- to high-altitude aircraft and cruise missiles traveling at speeds up to about MACH 2 (1,500 mph). To avoid detection, it was designed to be mobile and operate for only a few hours at one location.

The Patriot battery at Dhahran failed to track and intercept the Scud missile because of a software problem in the system's weapons control computer. This problem led to an inaccurate tracking calculation that became worse the longer the system operated. At the time of the incident, the battery had been operating continuously for over 100 hours. By then, the inaccuracy was serious enough to cause the system to look in the wrong place for the incoming Scud. The Patriot had never before been used to defend against Scud missiles nor was it expected to operate continuously for long periods of time. Two weeks before the incident, Army officials received Israeli data indicating some loss in accuracy after the system had been running for 8 consecutive hours. Consequently, Army officials modified the software to improve the system's accuracy. However, the modified software did not reach Dhahran until February 26, 1991—the day after the Scud incident (General Accounting Office Report on the Patriot Missile Incident 1992).

Much is made of the particular software error (a simple round-off error) that led to the failure of the Patriot missile defense system to protect American soldiers from the Scud missile. However, what this GAO report indicates, indirectly, is the importance

Figure 2.10 Patriot missile launcher. (Source: Redstone Arsenal historical information.)

of use cases. The Patriot missile system was designed to be mobile and operate for only a few hours at a time. There was no anticipation that it might be used for extended periods. Consequently, there were no use cases that covered the situation in which the failure occurred: continuous operation for over 100 hours. Without a use case, there were no functional or performance requirements to cover this situation. Without requirements, there was no test of whether such requirements would be met. Without a test, there was no way to discover the simple software error that led to the failure.

Prioritize Use Cases

Typically, a team of designers can come up with a very long list of possible use cases. The next step will involve describing in detail the behaviors that go along with these use cases. A team needs to limit the amount of effort it devotes to this activity. You will not describe behaviors for all of the use cases identified. Consequently, you should prioritize the use cases: Identify those cases with high priority (H), for which you must describe behaviors; cases with medium priority (M), for which you will describe behaviors if you have time; and cases with low priority (L), for which you

Table 2.8   Prioritized Use Cases for Toy Catapult

Use Cases	Priority
Child chooses toy	L
Child retrieves toy	L
Child plays with toy	H
Parent teaches child	L
Parent puts toy away	M
Parent entertains child	H

will not describe behaviors. The priorities should be based on how likely the use case will give rise to new and important functional requirements. The priorities in Table 2.8 reveal the fact that we believe the use cases of "child plays with toy" and "parent entertains child" will be the most fruitful to explore in detail. These two use cases have been given a high priority.

For more complex systems, you should budget about 10 high-priority use cases per team member. Each team member should be tasked with choosing a different set of 10 high-priority use cases and developing the behavioral descriptions for each use case in the set.

Describe Use Case Behaviors

The next step is to develop behavioral descriptions for each of the high-priority use cases. Table 2.9 illustrates a standardized spreadsheet template for presenting *use case behavioral descriptions* (Karas and Rhodes, 1987). Observe that the template has five sections:

1. *Use case name.* In this example, the use case is "child plays with toy."
2. *Initial conditions.* The initial conditions describe the state of the system and its environment prior to the onset of the behavior. You could have different behaviors for the same use case depending just on the initial state of the system. If this were the case, you might need to create a different spreadsheet for each behavior and initial condition. In this example, we assume the behavior starts with the toy catapult unloaded and unarmed.
3. *Behavior thread.* The central section of the spreadsheet describes the behavior as a thread—that is, as a sequence of events and activities. Each event or activity is listed on a separate row. Do not list two events or activities on the same row because it will be ambiguous as to which happens first. The exception is if you want to indicate that two activities occur concurrently in time. The events and activities are separated into three or more different columns depending on the entity with which the event or activity is most closely associated. By convention, we list the user or operator as the left-most column, the system itself as the second column, and all other external entities involved as the remaining columns. Typically, for a use case, the behavior thread begins with an event in the user or operator column. In this example, the behavior begins when the child pushes the receptacle into position. Observe that the language used to describe activities is different depending on what column the activity is listed in. If an activity is in the system column, it is a required activity of the system. We therefore describe it as a system requirement using the formal

Table 2.9   Behavioral Description of "Child Plays with Toy" Use Case

Child Plays with Toy		
Initial conditions		
1. System is in the unloaded state		
Operator (child)	**System (toy)**	**Projectile**
The child pushes the receptacle into position		
	The system shall detect the receptacle in proper position	
	The system shall secure the receptacle in position	
The child loads the receptacle with projectile		
		The projectile sits in the receptacle
	The receptacle shall hold the projectile	
The child triggers the release		
	The system shall detect the command to release from the child	
	The system shall eject the contents of the receptacle	
		The projectile flies through the air and lands some distance away
Ending conditions		
1. The toy is in the unloaded state		

Notes:

1. Assume the projectile is another toy (not the family pet gerbil!)

language of "the system shall... ." It may seem tedious to write out all of these events and activities in sentence form. You could more easily use verb phrases and abbreviations. However, it is better to provide full sentences at this stage, when you are in discovery mode and interested in the behavior, than to have to come back during a documentation phase when you will be less interested.

4. *Ending conditions.* The ending conditions describe the state of the system at the end of the behavior. This is a measure of success. It is considered a failed behavior if these ending conditions are not met at the end of the behavior. Note that we could have broken the behavior up into smaller threads. For example, we could have called this use case "child loads toy" and ended this thread with the activity "the system shall secure the receptacle in position." In this case, the ending condition would have been that "the system is in the armed state."

5. *Notes.* Because this is a process of discovery, ideas may occur to you as you write the behavioral description. Jot these notes down because they are a way to capture assumptions and concerns as they arise. For example, it was not until writing this behavioral description that the thought occurred to us that the projectile might be a pet!

It is typical to feel some anxiety at this stage because it is not clear at what level of detail you should be describing the behavior of a use case. Our advice is to begin at the level at which you or your team feels comfortable. Remember that whatever level of detail you choose, there comes a phase of the process where you will summarize the detail and end up with high-level, abstract functional requirements.

As you develop operational descriptions, you will be expanding the context and detail of the system in your mind. It is a good idea to record these details in the context matrix. Table 2.10 shows the context matrix for the toy catapult after thinking about the "child plays with toy" use case. The italicized entries in the matrix are new: they were not present in Table 2.6. Note that the entities "receptacle, lock, spring, and release" are internal entries. They may or may not be essential to the design but they are useful to describe the behavior.

The behavioral description of a second use case, "parent entertains child," can be found in Table 2.11. Rather than repeat the description of arming and loading the catapult, we begin this description from an initial condition of the catapult armed and loaded. The discovery from this behavioral analysis is how it would be useful to be able to trigger the catapult from a passing toy train. Note also that an additional column has been added to represent the train, an external entity. In the ending conditions, note that we require the train to continue running as part of the definition of success. The system would be considered a failure if it resulted in derailing the train.

Table 2.12 shows the revised context matrix, including the toy train as an external entity. The context matrix will continue to grow as you discover more about how the proposed product or service interacts with its environment.

Summarize Functional Requirements from Use Cases

By creating a collection of detailed behavioral descriptions, we have completed the "dive" aspect of this dive-and-surface approach. The next step is to "surface" and summarize the functional requirements. We use the drag-and-drop technique for the affinity process to perform this summarization (spreadsheet skill: drag and drop cells).

Table 2.10 Context Matrix after "Child Plays with Toy" Use Case

Is Related to	Child	Parent	Toy	Receptacle	Lock	Spring	Release	Projectile
Child			retrieves and plays with	pushes into position			triggers	places in receptacle
Parent	teaches and entertains		stores					
Toy	attracts			consists of	consists of	consists of	consists of	holds and launches
Receptacle			is part of					
Lock			is part of	secures in position		withstands force of	arms	
Spring			is part of	powers launch of				
Release			is part of		releases			
Projectile	amuses							

Notes: Italicized entries in the matrix are new: They were not present in Table 2.6. The system boundary is in bold.

Table 2.11   Behavioral Description of "Parent Entertains Child" Use Case

Parent Entertains Child with Train and Toy			
Initial conditions			
1. The toy is armed and loaded with projectile			
Operator (parent)	**System (toy)**	**Train**	**Projectile**
The parent places the toy beside the train track			
The parent starts the train			
		The train travels on the track	
		The train triggers the release as it passes the toy	
	The system shall detect the command to release from the train		
	The system shall eject the contents of the receptacle		
			The projectile flies through the air and lands some distance away
Ending conditions			
1. The toy is in the unloaded state			
2. The train continues to run on track			

Notes:

1. Assume the projectile is another toy (not the family pet gerbil!).

Table 2.12   Context Matrix after the "Parent Entertains Child" Use Case

Is Related to	Child	Parent	Toy	Receptacle	Lock	Spring	Release	Projectile	Toy Train
Child			retrieves and plays with	pushes into position			triggers	places in receptacle	*anticipates triggering event from*
Parent	teaches and entertains		stores						*aligns trajectory with toy release mechanism*
Toy	attracts			consists of	consists of	consists of	consists of		
Receptacle			is part of					holds and launches	
Lock			is part of	secures in position		withstands force of	arms		
Spring			is part of	powers launch of					
Release			is part of		releases				
Projectile	amuses								
Toy train	*amuses*						triggers		

Notes: Italicized entries in the matrix are new: they were not present in Table 2.10.

Table 2.13 Detailed Functional Requirements for Toy Catapult

Source
The system shall detect the receptacle in proper position
The system shall secure the receptacle in position
The receptacle shall hold the projectile
The system shall detect the command to release from the child
The system shall eject the contents of the receptacle
The system shall detect the command to release from the train

Once you have analyzed all of the high-priority use cases, all the functional requirements (requirements that appear in the system column of the behavioral descriptions) can be collected into an unordered list, as shown in Table 2.13.

Now, drag and drop the detailed functional requirements into grouping columns and write summary functional requirements at the head of each grouping column (Table 2.14).

Observe from this affinity process that if we had gone into excessive detail of operational description, we now have the opportunity to rewrite the functional requirements at a more simplified or abstract level.

Repeat for Secondary Use Cases

The *primary use cases* are the ones that naturally occur to us as we think about the way in which the system is going to be used. It is important to think also about ways in which the system can be

Table 2.14 Summarized Functional Requirements for Toy Catapult

System Shall Detect Its Receptacle in Proper Position	*System Shall Secure Its Receptacle in Position*	*System Shall Hold the Projectile in Its Receptacle*	*System Shall Detect the Command to Release from a Passing Toy (Lateral Direction) and Also from a Child or Falling Projectile (Vertical Direction)*	*System Shall Eject the Contents of Its Receptacle*
The system shall detect the receptacle in proper position	The system shall secure the receptacle in position	The receptacle shall hold the projectile	The system shall detect the command to release from the child	The system shall eject the contents of the receptacle
			The system shall detect the command to release from the train	

Table 2.15 Primary and Secondary Use Cases with Priorities

Use Case	Priority
Primary	
Child chooses toy	M
Child retrieves toy	L
Child plays with toy	H
Parent teaches child	L
Parent puts toy away	L
Parent entertains child with train and toy	H
Secondary	
Child releases armed toy near face of self or of another child	H
Child aims projectile at eyes of self or of another child	H
Child uses pet rodent as projectile	H
Child plays with toy repeatedly	H
Child drops or throws toy	H
Child leaves toy outside in bad weather	M

abused or the ways in which it can fail. It is frequently the case that some of the most important functional requirements come from a consideration of these cases. It is here that we can think through issues of safety and reliability. Table 2.15 lists both primary and *secondary use cases* associated with the toy catapult.

The detailed analysis of the behaviors for these secondary use cases is presented in Appendix C. This appendix also includes the most detailed context matrix from this analysis and the affinity table used to summarize the detailed secondary functional requirements.

Table 2.16 and Table 2.17 are the end result of this dive-and-surface approach to functional requirements analysis. Table 2.16 is a context matrix that focuses on the system and its external entities. The internal entities have been dropped from consideration. The internal identities will be considered in greater detail as the design progresses. Table 2.17 is a collection of all the high-level functional requirements that have been identified through this process. We refer to these as *originating requirements*—initial requirements identified by the owner as critical to the definition of the system. In conducting the functional requirements analysis, we have been operating on behalf of the owner. If we have identified an individual as owner, that individual would be called upon to approve or modify this list of originating requirements. Originating requirements are numbered to facilitate references to them. In particular, we will be deriving many detailed requirements from these originating requirements. We use reference numbers to trace derived requirements back to originating requirements. In the case of functional requirements, it is also useful to create an

Table 2.16 Summary Context Matrix for Toy Catapult

Is Related to	Child	Parent	Toy	Projectile	Other Moving Toys	Small Pets	Play Surfaces
Child	inappropriately aims toy at self or another		retrieves and plays with repeatedly; arms and loads; triggers release of	places in toy receptacle	anticipates triggering event from	inappropriately uses as projectile	drops or throws toy against
Parent	teaches, entertains, and trains in safety procedures		stores		aligns trajectory with toy release mechanism		
Toy	amuses but does not harm	warns of dangers and suggests appropriate age of child		holds and launches		does not harm	survives impact with
Projectile	amuses but does not harm						
Other moving toys	amuse		trigger				
Small pets			escape from or survive launch from				
Play surfaces							

Note: The system boundary is indicated by the bold border.

Table 2.17 Draft of Originating Requirements

Index	Originating Requirements	Abstract Function Name
OR.1	The system shall detect its receptacle in proper position.	Detect
OR.2	The system shall secure its receptacle in position.	Arm
OR.3	The system shall hold the projectile in its receptacle.	Load
OR.4	The system shall detect the command to release from a passing toy (lateral direction) and also from a child or falling projectile (vertical direction).	Trigger
OR.5	The system shall eject the contents of its receptacle.	Eject
OR.6	The system shall notify the parent of the dangers and inherent risks of the system.	Warn
OR.7	The system shall suggest to the parent the appropriate age of the child to use the system.	Suggest age
OR.8	The moving parts of the system shall be prevented from striking the face of a child with sufficient force to cause a bruise.	Protect face
OR.9	The system shall not launch projectiles from the child's domain with sufficient force to cause damage to the child or the projectile.	Constrain launch force
OR.10	The system shall enable the escape of a pet rodent from the receptacle.	Enable pet escape
OR.11	The system shall successfully complete each cycle of use for many cycles.	Repeat
OR.12	The system shall survive a collision with a hard surface in working order.	Survive collision
OR.13	Verification of all requirements shall be conducted in such a way as to preserve the health of human and animal subjects.	Test safely

abstract function name to refer to the requirement. Table 2.17 illustrates this with a list of *function names* to be used as shorthand for the functional requirements.

Finalize Requirements

Before accepting a collection of originating requirements as final, we should check to make sure the requirements are not ambiguous, overly restrictive, or contradictory. It is possible later in the design cycle to come back and clarify or challenge the validity of originating requirements, but the later these challenges occur, the more difficult they are to resolve. A great deal of design effort can be wasted if the originating requirements are flawed in some way.

Reviewing the originating requirements of Table 2.17 for the toy catapult, we can spot a number of potential problems. In Table 2.18, we list a number of *issues* and their *resolution*. The resolution appears as a rewritten requirement. Because the owner is responsible for originating requirements, we would need to get approval for these changes from the owner before proceeding with the design. Table 2.19 lists the finalized originating requirements.

Summary

In this chapter, we began with a product concept—a simple idea of a product—sketched it, and used a context diagram and context matrix to identify the entities that made up the product (internal entities) and the entities related to the product (external entities). This helped to limit the scope of our design. We identified three different points of view from which to study our product: the owner, the customer, and the user, and we realized that the product can look quite different depending on the point of view. We expressed our goals for the design by means of a mission statement. We made an effort to study the context in which the product would be used, noting that we can play the role of anthropologist, apprentice, or partner as we conduct such studies. In particular, we gathered detailed comments from customers providing their attitudes about the type of product we were contemplating. We used a two-step affinity process to organize these comments into categories and then to give descriptions of these categories. In this way, we summarized the voice of the customer.

Using brainstorming techniques, we generated a list of primary and secondary use cases for the product. The primary use cases were normal ways in which users would interact with the product. The secondary use cases reflected abnormal cases or safety concerns. We then prioritized the use cases, and for the high-priority use cases, we developed detailed use case behavioral descriptions. Each behavior consisted of a thread of activities from a stimulus to a response or set of responses. We wrote activities that the product performed in the formal fashion of a functional requirement, expressed as "the system shall" statements.

As we developed these behavioral descriptions, we expanded our context matrix to reflect the relevant entities being discovered. We collected all the detailed functional requirements from these behavioral descriptions and used the affinity process again to summarize them as more abstract originating requirements. We also simplified the context matrix to focus on the external entities and their relationships to our product. We considered issues with the functional requirements and resolved them by writing clearer finalized originating requirements.

By the end of the chapter, we had created a product concept diagram, a context matrix, a mission statement, a table of product objectives that summarized the voice of the customer, and a list of finalized originating functional requirements. It is a good start.

Discussion

1. Identify other design situations, such as the infant mobile example, in which the user's viewpoint is often forgotten.
2. Suggest a process by which, had it been implemented, the software bug in the Patriot missile might have been discovered sooner.

Table 2.18 Resolving Issues with Originating Requirements

Index	Originating Requirements	Function Name	Issues	Resolution
OR.1	The system shall detect its receptacle in proper position	Detect	Issue: This assumes a particular spring-based design. We can imagine a remote-controlled battery-operated toy. Is there a more general way to state the requirement?	The system shall detect the command to arm
OR.2	The system shall secure its receptacle in position	Arm		The system shall secure its receptacle in the armed position. The force required to arm the system may be supplied by the system (e.g., battery power) or by the child
OR.3	The system shall hold the projectile in its receptacle	Load		
OR.4	The system shall detect the command to release from a passing toy (lateral direction) and also from a child or falling projectile (vertical direction)	Trigger		
OR.5	The system shall eject the contents of its receptacle	Eject	Issue: The restriction of OR.9 may prevent the launch of all possible contents. What contents are expected?	The system shall be capable of ejecting lightweight balls and dolls from its receptacle
OR.6	The system shall notify the parent of the dangers and inherent risks of the system	Warn		

OR.7	The system shall suggest to the parent the appropriate age of the child to use the system.	Suggest age		
OR.8	The moving parts of the system shall be prevented from striking the face of a child with sufficient force to cause a bruise.	Protect face		
OR.9	The system shall not launch projectiles from the child's domain with sufficient force to cause damage to the child or the projectile.	Constrain launch force		
OR.10	The system shall enable the escape of a pet rodent from the receptacle.	Enable pet escape		
OR.11	The system shall successfully complete each cycle of use for many cycles.	Repeat		
OR.12	The system shall survive a collision with a hard surface in working order.	Survive collision	Issue: Do we mean that the hard surface must be in working order?	The system shall survive a collision with a hard surface and remain in working order.
OR.13	Verification of all requirements shall be conducted in such a way as to preserve the health of human and animal subjects.	Test safely		

Table 2.19 Finalized Originating Requirements

Index	Originating Requirements	Abstract Function Name
OR.1	The system shall detect the command to arm.	Detect
OR.2	The system shall secure its receptacle in the armed position. The force required to arm the system may be supplied by the system (e.g., battery power) or by the child.	Arm
OR.3	The system shall hold the projectile in its receptacle.	Load
OR.4	The system shall detect the command to release from a passing toy (lateral direction) and also from a child or falling projectile (vertical direction).	Trigger
OR.5	The system shall be capable of ejecting lightweight balls and dolls from its receptacle.	Eject
OR.6	The system shall notify the parent of the dangers and inherent risks of the system.	Warn
OR.7	The system shall suggest to the parent the appropriate age of the child to use the system.	Suggest age
OR.8	The moving parts of the system shall be prevented from striking the face of a child with sufficient force to cause a bruise.	Protect face
OR.9	The system shall not launch projectiles from the child's domain with sufficient force to cause damage to the child or the projectile.	Constrain launch force
OR.10	The system shall enable the escape of a pet rodent from the receptacle.	Enable pet escape
OR.11	The system shall successfully complete each cycle of use for many cycles.	Repeat
OR.12	The system shall survive a collision with a hard surface and remain in working order.	Survive collision
OR.13	Verification of all requirements shall be conducted in such a way as to preserve the health of human and animal subjects.	Test safely

Exercises

1. Use Google Images™ to search for "product concept sketch." Submit four product concept sketches (copy and paste into your assignment document and attribute the source) fulfilling the following criteria:
   a. a sketch that visually excites you;
   b. an annotated sketch (as in Figure 2.3);
   c. a sketch that conveys the product concept extremely well (information-rich even if it has little or no text); and
   d. a sketch that you could probably create yourself in less than 45 minutes. Remember that the primary goal is communication; artistic quality is secondary.

2. Use Google Images to search for "product concept diagram." As you will see, they come in many varieties. Submit three product concept diagrams (copy and paste into your assignment document and attribute the source) fulfilling the following criteria:
   a. diagram with nodes and arcs that does a good job of conveying the product concept—nodes of any type (ovals, squares, pictures, or icons) and arcs of any style;
   b. a diagram with nodes and arcs in which the arcs are labeled with phrases that describe the relationships between nodes; and
   c. a diagram with nodes and arcs in which the arcs are not labeled, so it is very difficult to figure out the relationships.

3. Create an annotated product sketch of your own to illustrate some product concept that interests you. You may scan a drawing of your own or use a drawing or paint program or use clip-art—any technique as long as the composition is uniquely yours. The sketch with the title and annotations should be sufficient for any reader to understand your concept.

4. Describe your concept using complete sentences. Your concept sketch will be compared with your verbal description to see how well the sketch communicates the idea.

For the next two exercises, you may continue to develop your own product idea or you may use one of the product design challenges, such as the BathBot (the bathroom-cleaning robot challenge located in Appendix B). It will be a little more work to do your own idea. Just make sure that you go to a similar level of detail with your idea as we have suggested for the BathBot idea.

5. Download the file "ContextDiagrams.xls" from the text Web site. On the sheet labeled "BathBot," complete the concept diagram for the BathBot robot by drawing nondirectional rectilinear arcs between the BathBot central node and the outside entities (or between outside entities). (You can find rectilinear arc drawing tools in the AutoShapes–Connectors menu of MS Excel. You may need to enable the drawing toolbar under View–Toolbars–Drawing.) Label each arc with a verb phrase near the node that would form the subject of the sentence. Examples of relationships that might occur to you are "maintenance technician services BathBot II" and "maintenance technician orders replacement parts from replacement parts supply."

6. Complete the context matrix on the sheet labeled "BathBot" in the same file using the relationships you captured in the previous question. Size the rows and columns and use word wrap formatting so that all of your text is readable.

7. Summarize the customer comments in the file "BathBotCustomerComments.xls" using the affinity process and the Excel drag-and-drop technique. That is, drag each customer comment and place it in a group column with other similar customer comments. Replace the column heading group name with a sentence expressing what the customer is asking for in that column. Use the customer's language (such as "make it economical"). Try not to have too many different groups. Take the groups and further summarize them to get a small number of attributes that the customer is looking for (at most six). On a separate sheet in the workbook, create a table that summarizes the "Voice of the BathBot Customer" in a format similar to Table 2.5 in the text. (Suggestion: download and study the file "Toy 02 Customer Comments.xls" to see how the affinity process was applied, step by step, to the toy example.)

8. Download and use the spreadsheet template "FunctionalRequirementsAnalysis.xls" to conduct a simplified functional requirements analysis for either the bathroom-cleaning robot or for the product concept of your choice. By "simplified," we mean that you will perform all the basic steps of the process, but you will consider only a very small number of use cases. (Suggestion: begin by downloading and studying the file "Toy 01 Functional Requirements.xls" to see all the steps applied.)

    a. Start with a simplified context matrix, like Table 2.6.
    b. Restrict yourself to three primary use cases and no secondary use cases. Choose cases that are similar (but not identical) so that there is a good chance that you will see the same functionality from different perspectives. For example, cleaning the toilet is likely to have almost exactly the same functional requirements as cleaning the sink except for what to do about the toilet seat and sanitary issues. Coping with the toilet seat will likely give rise to new functional requirements. Three suggested use cases are to
        i. prepare to start BathBot (initial conditions assume BathBot is in the bathroom but is not connected to power, water, or drain);
        ii. clean sink (from initial conditions of being connected to power, water, and drain); and
        iii. clean toilet (from initial conditions of being connected to power, water, and drain, and lid down).;
    c. Describe the behavior of these use cases using the format of Table 2.9.
    d. Expand the context matrix as in Table 2.12. Show the internal and external entities you have discovered.
    e. Collect the functional requirements and summarize them, as in Table 2.14. At this level of abstraction, you should not be talking about specific bathroom fixtures. These should be generic functions concerning generic bathroom fixtures.
    f. (Skip the analysis of secondary use cases.) Summarize the context matrix by eliminating the internal entities, as in Table 2.16.
    g. Draft the list of originating requirements, as in Table 2.17.

References

Beyer, H., and K. Holtzblatt. 1998. *Contextual design: Designing customer-centered systems.* San Francisco: Morgan Kaufmann Publishers.

Condit, P., J. Guyette, and R. Albrecht. 1990. B777 objectives. Handwritten note. The Boeing Corporation.

Corson, Bruce. 2009. Personal communication.

Edwards, B. 1999. *The new drawing on the right side of the brain.* New York: Jeremy P. Tarcher/Putnam.

Karas, L., and D. Rhodes 1987. Systems engineering technique. NATO Advisory Group for Aerospace Research & Development Conference Proceedings no. 417.

McKinzie, G. A. 1996. How United and Boeing worked together to design and build the 777 airplane. *National Productivity Review* (Winter) pp. 7–14.

Norris, G. 1995. Boeing's seventh wonder. *IEEE Spectrum* (October): 20–23.

Rosenbloom, S. 2007. My dad, the inventor. *New York Times* Aug. 16, 2007, Fashion & Style.

U.S. GAO. 1992. Patriot missile software problem. United States General Accounting Office Report GAO/IMTEC-92-26.

Chapter 3

Measure the Need
and Set Targets

All designs are a compromise on the designer's attempt to combine certain desirable features without sacrificing too much safety, comfort, and cost.

L. Francis Herreshoff, 1943 (as quoted in Caswell 2008)

Introduction

In this chapter, we focus on measurement and targets. It is our goal to take the customer product objectives that were identified from our analysis of customer comments in the previous chapter, and (1) find ways to measure how well any product can satisfy those objectives, (2) find ways to measure how important these different product objectives are to potential customers, and (3) set technical performance targets for our own design concept that will make our design better at satisfying those product objectives than any of the competing products. You can appreciate that going from vague product objectives, stated in English, to detailed performance specifications, specified in engineering terms, is a valuable skill. This chapter offers a road map for acquiring that skill.

The process is organized into three major steps and the chapter follows this organization:

1. Measure the need.
2. Translate the need into technical requirements.
3. Identify the customer value proposition.

Measure the Need

A *metric* is a quantitative measure of the degree to which a system, component, or process possesses a given attribute. A *measure of effectiveness* is a metric used to measure results achieved against those specified in a mission statement. Recall our mission statement for the toy catapult example: "Our mission in this example is to design a toy catapult to amuse a 4-year-old child safely, creating an opportunity for parent and child to play together in a way that delights both parent and child."

The problem we face in this section is to develop measures of effectiveness by which we may judge our success (or the success of our competitors) in this mission. Our goal is to quantify the customer needs, rank them in importance, and develop measurement schemes that permit us to assess how well our design, as well as competitive designs, meet the needs of our customers.

The primary motivation for collecting data and for measuring effectiveness is a clinical view of the design process. The major advances in the treatment of disease in the past 150 years have come about from clinical medicine: the recording of treatments and their results. Clinical studies have enabled doctors to develop a statistical view of treatment effectiveness. If you view design as a one-time effort, then the task of collecting data and measuring the effectiveness of your design will likely seem to be a wasted effort. But, with this mentality, you will approach every design project as a one-time effort. Project after project will come and go, and you will have no quantitative basis to state that your design performance is improving. On the other hand, a clinical view of the design process is that each cycle of design is an opportunity to learn something. By collecting data on each cycle, particularly data that relate to measures of effectiveness, you will have a solid experimental basis on which to improve your techniques.

A second motivation for developing quantitative measures of effectiveness is to convert problems in design into problems of engineering. Whether or not you consider yourself to be an engineer, the ability to pose design problems as engineering problems is a valuable skill.

The Relaxacizor

The notice below concerns a product, the Relaxacizor (Figures 3.1A and 3.1B), so detrimental to public health that even secondhand sales of the device were banned and owners were advised to destroy or disable them. The device was sold as an exercise machine because it could cause muscles to jump in response to electrical shocks. It was determined to have no exercise value but instead to have the potential for considerable bodily harm. The story provides an example of an unbalanced design: a design that neglects the full range of concerns for a product's use.

U.S. DEPARTMENT OF HEALTH, EDUCATION, AND WELFARE
PUBLIC HEALTH SERVICE
FOOD AND DRUG ADMINISTRATION
WASHINGTON, D.C. 20204
71-1
FOR RELEASE:
WEDNESDAY, JANUARY 13, 1971

The Food and Drug Administration warned today that the sale of second-hand relaxacizors is illegal.

The warning stemmed from reports that owners of the electrical devices are attempting to dispose of them by offering them for sale in classified advertisements.

The devices provide electrical shocks to the body through contact pads. They were declared dangerous to health in a California court ruling last April against Relaxacizor, Inc., the distributor.

In his decision, Judge William P. Gray said the devices could cause miscarriages and could aggravate many pre-existing medical conditions, including hernia, ulcers, varicose veins and epilepsy.

More than 400,000 units have been sold for exercise and reducing. After Judge Gray's decision, many concerned owners wrote to the distributor requesting a refund. In seeking to allay their fears, the firm said it had filed an appeal and was confident the ruling would be reversed.

The appeal was dismissed last November 24, by agreement between the firm and the Government, thus ending the case.

In recent weeks, FDA has supplied posters to all post offices warning against use of the devices.

In its warning today to both sellers and prospective purchasers of the devices, the FDA said such sales are in violation of the Food, Drug and Cosmetic Act and the devices are subject to seizure.

Figure 3.1A   The Relaxacizor instruction manual cover. (Courtesy of Bobby Beeman.)

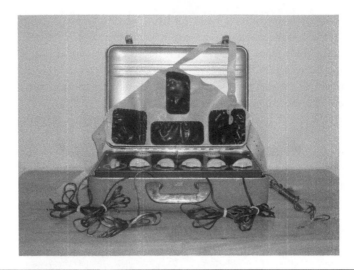

Figure 3.1B   The Relaxacizor.

The agency recommended that owners of the device either destroy them or render them inoperable to avoid any possibility of harm to unsuspecting users.

U.S. Dept. of Health, Education and Welfare, 1971
Museum of Questionable Medical Devices, March 2000

Determine Measures of Effectiveness

Data collection is often conducted in a haphazard way. It is a common discovery that the data collected in a study cannot be used in any meaningful way. This can be traced to the way in which the data collection study was designed. The problem is that people tasked with collecting data often use traditional predefined metrics because they assume that what is traditional must be useful.

Suppose a product development team considers removing a certain feature or functionality from an existing product, such as a speed dial button from a handheld phone. That feature may be expensive or it may increase the complexity of the design. Someone on the team calls for a customer study to answer the question. A customer study team is commissioned. The customer study team collects survey data on where the product was purchased and on the age, sex, marital status, and income of the customer who purchased the product. They may also ask whether the product was purchased for business, educational, or entertainment reasons. As a consumer, you have likely filled out many surveys like this yourself. You can see immediately, however, that these data are only marginally useful in answering the specific design question that was posed originally. There is no connection between the goal of the study and the data collected. The product development team will be frustrated by the lack of meaningful results from the customer study.

It can be difficult to think of quantitative ways to measure effectiveness. The mission statement for the toy catapult includes the phrase "delights both parent and child." Exactly how can "delight" be measured?

The *goal–question–metric (GQM) method* is an approach to data collection that is useful anywhere that metrics are required to assess the satisfaction of design goals. Dr. Victor Basili and his colleagues at the NASA Software Engineering Laboratory (SEL) developed the GQM approach during the 1980s (Basili and Weiss 1984). It is now a widely used approach, especially in software development and maintenance.

There are four steps in the GQM method of determining measures of effectiveness:

1. Identify the goals of the measurement.
2. Generate questions that define the goals in a quantitative way.
3. Specify the measures needed to answer the questions.
4. Develop mechanisms to collect the data.

Observe that the method is organized into a conceptual level (identify goals), an operational level (formulate questions), a quantitative level (specify metrics), and a procedural level (data collection). We illustrate the steps with reference to the toy catapult example.

The Goal–Question–Metric Method Applied to the Toy Catapult

Identify the Goals of the Measurement

Step 1 of the goal–question–metric method is to identify the goals of the measurement. We have already performed that step during our phase of customer analysis. Recall how we represented the "voice of the customer" based on affinity analysis of customer comments. Table 3.1 repeats the results of that analysis.

To keep the analysis simple, we focus only on the high-level product objectives, as extracted in Table 3.2.

Table 3.1 The Voice of the Customer

Make the toy fun for both parent and child	Make a toy that interests my child
	Make a toy that involves me, the parent
	Don't make the toy too complex
Make the toy appealing for both parent and child	Make a toy with personality
	Make a toy that is pleasing to the eye and to the touch
	Don't make a toy with annoying noise
Make the toy safe	Don't let the toy injure my child
Make the toy easy to put away	Make the toy easy to put away (and find again)
	Make a toy that is easy to store
Make the toy playable for a long time	Make the toy reliable
	Avoid making a battery-powered toy

Table 3.2   Product Objectives

Make the toy fun for both parent and child
Make the toy appealing for both parent and child
Make the toy safe
Make the toy easy to put away
Make the toy playable for a long time

Before leaving this step of identifying goals, it is useful to decompose the goal into its constituent parts. This is a check to ensure the clarity of the goal. The constituent parts of a goal are as follows:

1. The object or process under measure. This typically will be the system (product or process) at the center of your context matrix or one of the relations you discovered with outside entities. For example, you may focus on the repair process that occurs when your product breaks.
2. The purpose of the measurement activity (understanding, design, control, or improvement). That is, is the purpose to understand the behavior of the object; to control its behavior, keeping it within certain limits; to improve the object or process from some perspective; or to design a new object or process that achieves some objective you would like to measure?
3. The quality focus of the object or process on which the measurement focuses. That is, name the attribute in which you are most interested.
4. The perspective or viewpoint of the measurement. For example, measuring comfort in an automobile takes on different meanings depending on whether you are the driver, a front-seat passenger, or a backseat passenger.
5. The context or environment in which the measurement takes place. Laboratory environments, for example, are very different from workplace environments. Products that work well in controlled laboratories sometimes fail in actual use.

Table 3.3 analyzes the product objectives for the toy catapult. We now have names for the different quality attributes (fun, appeal, safety, storability, and reliability). This approach has also forced us to be more explicit about the context of the measurement. We note that the environment we picture for the product evaluation is a single-family home. This is to distinguish it from, say, a behavioral research facility. The environment of a single-family home was implicit in our customer analysis, but this line of thinking has forced us to make it explicit.

Refine the Goals with Questions

The step that is often missing in data collection studies is the formulation of the questions that the data are intended to help answer. Without this step, you will likely end up collecting data related to the goals of measurement but insufficient to answer any questions. Asking the questions first, before the data are collected, is the best assurance that the data will yield meaningful answers.

The questions must take the goals to a greater level of specificity than the original statement of the goals. Simply rephrasing the goal as a question is not useful. To be useful, the question must identify an aspect of the goal that can be tested in a quantitative fashion.

Table 3.4 lists some possible questions that could be used to refine the product objectives for the toy catapult. Observe that the questions are much more quantitatively oriented than the original

Table 3.3 Analyzing Product Objectives

	Analyze (The Object or Process under Measure)	**For the Purpose of** (Understanding, Designing, Controlling, or Improving the Object)	**With Respect to** (The Quality Focus of the Object that the Measurement Focuses On)	**From the Perspective of** (The People that Measure the Object or who Value the Attribute)	**In the Context of** (The Environment in which the Measurement Takes Place)
Make the toy fun for both parent and child	toy	design	fun	parent, child	single-family home
Make the toy appealing for both parent and child	toy	design	appeal	parent, child	single-family home
Make the toy safe	toy	design	safety	child	single-family home
Make the toy easy to put away	toy	design	storability	parent	single-family home
Make the toy playable for a long time	toy	design	reliability	parent	single-family home

goals. For example, rather than trying to directly measure the fun of the toy, we ask whether it is played with frequently or for long periods of time. The safety goal has been translated into a simple assessment of the appropriate child age by the parent. You recall that a functional analysis of safety concerns in the previous chapter gave rise to an extensive list of requirements.

Specify the Metrics

Equipped as we are with a detailed set of questions, it is possible now to design metrics and ways of collecting data that will answer the questions. If we can show, for example, that a toy of our design can be picked up and put away faster than other toys and that it takes up less space than other toys, then we have a quantitative basis for saying that our design has met the goal of being easy to put away.

In designing metrics, we must trade off the quality of information gained from the data with the difficulty of collecting the data. We suggest making two passes at the problem. In the first

Table 3.4   Refining Goals with Questions

Goal	Questions
Make the toy fun for the parent	Does the parent play with the child and this toy frequently?
	Does the parent play with the child and this toy for long periods of time?
Make the toy fun for the child	Does the child play with the toy frequently?
	Does the child play with the toy for long periods of time?
Make the toy appealing to the parent	Does the parent like the look and feel of the toy?
Make the toy appealing to the child	Does the child have a special name for the toy?
	Is the child likely to select this toy first from among many toys?
Make the toy safe	What is the recommended minimum age for a child to be old enough to play with this toy safely and responsibly?
Make the toy easy to put away	How long does it take the parent to gather the toy together (if there are multiple pieces)?
	How long does it take the parent to shelve the toy?
	How much space does the toy take up on the shelf?
	How much storage space for other toys does this toy provide?
Make the toy playable for a long time	How long does the toy last until its first maintenance is required (e.g., replacing batteries)?
	How long does the toy last until it is materially damaged?
	How long does the toy last until it is unplayable in its original form?
	How long does the toy last until it is discarded?

pass, write down an *ideal metric*—that is, a measurement that would directly answer the question being posed. Ignore, for this pass, the difficulty of collecting the data. Just imagine what it takes to answer the question well. In the second pass, consider the difficulty and expense of collecting the data and propose an *approximate metric*—one that is likely to be highly correlated with the ideal metric but is more practical, given your budget for data collection. By separating the issues of quality and practicality, you will likely find that the overall process of defining metrics goes much faster. Furthermore, if you are challenged to justify the choice of an approximating metric, you can focus the discussion on how correlated it is likely to be with the ideal.

Table 3.5 reveals how we developed metrics to answer the questions raised for the toy catapult. Observe the trade-off between quality of information and cost of data collection that is evident in the design of approximate metrics. For example, to answer the question, "Does the child play with the toy for long periods of time?" the ideal metric is the fraction of total playtime that the child spends with this toy. This ideal metric would require constant monitoring of the child's playtime. It is unrealistic to ask a parent to collect these types of data (if the parent is our intended data collector). A simpler data collection scheme is simply to ask the parent to walk into the play area at random times and make a note of the toys with which the child is playing. The corresponding metric would be the fraction of random visits in which the child is playing with this particular toy.

As another example, the ideal way to answer how long it takes the parent to gather the toy together is to measure this time on a daily basis and then average the times. That, again, is an excessive data collection burden. A simpler metric would be the number of pieces that make up the toy. This can be determined by studying the design of the toy and is likely to be correlated highly with the time required to gather the pieces.

Develop Data Collection Methods

The final step in the GQM method is to develop data collection methods to calculate the metrics. As you think through these data collection methods, you may realize the difficulty of data collection and return to the previous step to specify more cost-effective metrics.

Table 3.6 summarizes the data collection methods we have in mind for a study of how successful different toys are at meeting the goals of fun, appeal, safety, storability, and reliability. We see the need for two different types of data collection tools: a questionnaire that requires only one-time answers to specific questions for each toy, and a data collection grid (i.e., a table) that would be used to record information from multiple playtime sessions.

Table 3.7 is a sample *data collection grid*. This one is designed to capture how frequently (and perhaps how long) different toys are being played with. The idea is that the parent would observe the child periodically and, perhaps, at random intervals, when the child is playing and would note on this grid the toy with which the child is playing. Every time the parent notices a different toy in use, he or she would add a column to track it, writing the type of toy at the head of the column. The form is designed to require only a minimal amount of the parent's time to fill out. Each entry should require only a few seconds to write down: a quick estimate of the time since the last entry was made—using abbreviations such as "dd" to represent several days and "mm" to represent several minutes and a check mark in the column of the toy in play. After a couple of weeks of data collection and several hundred observations, we should have a good basis to identify which toys are the child's current favorites. If numerous entries are separated by only a few minutes, then we could also identify toys that are played with for a long time.

Similar data collection grids can be designed for capturing what toys are used when the parent plays with the child and to measure the appeal the different toys hold for the child.

Table 3.5   Defining Metrics to Answer Questions

Goal	Questions	Ideal Metric	Approximate Metric
Make the toy fun for the parent	Does the parent play with the child and this toy frequently?	Fraction of days in which this toy was played with jointly (both parent and child) at least once	(No substitute)
	Does the parent play with the child and this toy for long periods of time?	Fraction of total joint playtime spent with this toy	(No substitute)
Make the toy fun for the child	Does the child play with the toy frequently?	Fraction of days in which this toy was played with at least once	(No substitute)
	Does the child play with the toy for long periods of time?	Fraction of total playtime spent with this toy	Fraction of random visits to playroom in which child is found to be playing with this toy
Make the toy appealing to the parent	Does the parent like the look and feel of the toy?	Subjective score relative to other toys; score ranges from "much worse than other toys" to "much better than other toys"	(No substitute)
Make the toy appealing to the child	Does the child have a special name for the toy?	Parental answer: yes/no	(No substitute)
	Is the child likely to select this toy first from among many toys?	Fraction of all play sessions in which this is the first toy played with	Fraction of times this toy is first toy played with after the child's nap
Make the toy safe	What is the recommended minimum age for a child to be old enough to play with this toy safely and responsibly?	Recommended minimum age	(No substitute)

Table 3.5   Defining Metrics to Answer Questions (Continued)

Goal	Questions	Ideal Metric	Approximate Metric
Make the toy easy to put away	How long does it take the parent to gather the toy together (if there are multiple pieces)?	Seconds to collect scattered toy	Number of pieces
	How long does it take the parent to shelve the toy?	Seconds to assemble and shelve collected toy	Subjective score relative to other toys; score ranges from "much worse than other toys" to "much better than other toys"
	How much space does the toy take up on the shelf?	Volume on shelf unusable by other toys (cubic inches)	Volume of smallest cube containing toy (cubic inches)
	How much storage space for other toys does this toy provide?	Usable storage space (cubic inches)	Theoretical storage space (cubic inches)
Make the toy playable for a long time	How long does the toy last until its first maintenance is required (e.g., replacing batteries)?	Hours of play until first maintenance	Weeks from first play to first maintenance
	How long does the toy last until it is materially damaged?	Hours of play until first breakage	Weeks from first play to first breakage
	How long does the toy last until it is unplayable in its original form?	Hours of play until failure	Weeks from first play to failure
	How long does the toy last until it is discarded?	Total hours of play until discard	Months from first play until discard

Observe a few principles in designing data collection forms:

1. Design data collection forms to fit on a single sheet of paper.
2. Strive to minimize the time and effort required for someone to make a data entry. Imagine that each character that must be written down requires 1 second. Simplify your data requirements so that an entry requires only a few seconds.

Table 3.6 Developing Data Collection Methods

Goal	Questions	Ideal Metric	Approximate Metric	Data Collection Method
Make the toy fun for the parent	Does the parent play with the child and this toy frequently?	Fraction of days in which this toy was played with jointly (parent and child) at least once	(No substitute)	Data collection grid
	Does the parent play with the child and this toy for long periods of time?	Fraction of total joint playtime spent with this toy	(No substitute)	Data collection grid
Make the toy fun for the child	Does the child play with the toy frequently?	Fraction of days in which this toy was played with at least once	(No substitute)	Data collection grid (events by day)
	Does the child play with the toy for long periods of time?	Fraction of total playtime spent with this toy	Fraction of random visits to playroom in which child is found to be playing with this toy	Data collection grid (events by visit)
Make the toy appealing to the parent	Does the parent like the look and feel of the toy?	Subjective score relative to other toys; score ranges from "much worse than other toys" to "much better than other toys"	(No substitute)	Questionnaire
Make the toy appealing to the child	Does the child have a special name for the toy?	Parental answer: yes/no	(No substitute)	Questionnaire
	Is the child likely to select this toy first from among many toys?	Fraction of all play sessions in which this is the first toy played with	Fraction of times this toy is first toy played with after the child's nap	Data collection grid (events by days)
Make the toy safe	What is the recommended minimum age for a child to be old enough to play with this toy safely and responsibly?	Recommended minimum age	(No substitute)	Questionnaire

Table 3.6 Developing Data Collection Methods (Continued)

Goal	Questions	Ideal Metric	Approximate Metric	Data Collection Method
Make the toy easy to put away	How long does it take the parent to gather the toy together (if there are multiple pieces)?	Seconds to collect scattered toy	Number of pieces	Questionnaire
	How long does it take the parent to shelve the toy?	Seconds to assemble and shelve collected toy	Subjective score relative to other toys; score ranges from "much worse than other toys" to "much better than other toys"	Questionnaire
	How much space does the toy take up on the shelf?	Volume on shelf unusable by other toys (cubic inches)	Volume of smallest cube containing toy (cubic inches)	Questionnaire (dimensions)
	How much storage space for other toys does this toy provide?	Usable storage space (cubic inches)	Theoretical storage space (cubic inches)	Questionnaire (dimensions)
Make the toy playable for a long time	How long does the toy last until its first maintenance is required (e.g., replacing batteries)?	Hours of play until first maintenance	Weeks from first play to first maintenance	Questionnaire
	How long does the toy last until it is materially damaged?	Hours of play until first breakage	Weeks from first play to first breakage	Questionnaire
	How long does the toy last until it is unplayable in its original form?	Hours of play until failure	Weeks from first play to failure	Questionnaire
	How long does the toy last until it is discarded?	Total hours of play until discard	Months from first play until discard	Questionnaire

Table 3.7 Sample Data Collection Grid: Frequency of Toy Usage

Child's Name:					Observer's Name:				
Observation Period: From: / /					To: / /				
Locale:									
		Toy being played with when checked							
Check no.	Approximate time since last check (e.g., several days, a day, a few hours, or a few minutes, abbreviated as "dd," "d," "hh," and "mm," respectively)								
1									
2									
3									
4									
5									
6									
7									
8									
9									
10									

3. Add identification data to the form, such as who collected the data, over what period of time, and in what location or circumstance.
4. Test your technique. After you start to collect data, you will likely discover problems in your technique. Be prepared to redesign the form.

Repeat for Secondary Goals

During the initial customer analysis phase of the project, it is possible to miss concerns that should be considered. This is a good opportunity to review the different objectives on which you are measuring success and to attach secondary goals to the project if they are seen as important to mission success. For example, in our study of parental comments about toys, there was no mention of how difficult it can be to repair a toy once it breaks. Depending on the materials used, some toys can be glued back together easily if they break. Others break in ways that are irreparable. Similarly, the cost and effort to make the toy are definitely a concern for the owner of the system, whom we identified to be me, the grandfather.

For the toy catapult, Table 3.8 lists the secondary goals identified and analyzes these goals into their constituent parts. As in the case of the primary goals, the analysis has forced us to be specific

Table 3.8 Secondary Goals Analyzed

	Analyze (The Object or Process under Measure)	*For the Purpose of (Understanding, Designing, Controlling, or Improving the Object)*	*With Respect to (The Quality Focus of the Object that the Measurement Focuses On)*	*From the Perspective of (The People that Measure the Object or Who Value the Attribute)*	*In the Context of (The Environment in which the Measurement Takes Place)*
Make the toy easy to repair	toy	design	repairability	parent	single-family home
Make the toy affordable	toy	design	affordability	grandparent	hobbyist shop

about the environment of the measurement. When evaluating affordability, we need to consider the cost and effort of creating a toy using only hobbyist tools and equipment, rather than a full-scale manufacturing plant.

Table 3.9 follows the goal–question–metric method through to completion for these secondary goals.

Weight the Product Objectives

Until now, we have not been forced to compare product objectives. We have been concerned with finding out what aspects of the product are important, capturing them, and finding ways to measure our success in those dimensions. A time will come in the design process when we must make decisions that trade off performance in one dimension against performance in another dimension. It is unrealistic to expect that we can excel in all dimensions simultaneously. If you want a vehicle that is highly maneuverable in traffic, you will design a motorcycle. If you want to stay dry and

Table 3.9 Secondary Goals with Questions, Metrics, and Methods

Goal	*Questions*	*Ideal Metric*	*Approximate Metric*	*Method*
Make the toy easy to repair	How long does it take to replace a broken latch?	Mean time to repair (minutes)	(No substitute)	Estimate from design
Make the toy affordable	What is the cost of materials?	Dollars of materials required	(No substitute)	Estimate from design
	How long does it take to build?	Hours of construction required	(No substitute)	Estimate from design

carry on a conversation while you wait in traffic, you will design a sedan. Trade-offs are inevitable. It is said that "all design is a compromise."

Because we will be forced eventually to make design choices, this is an appropriate time to reflect on which product objectives are most important to us. At this stage, we have no vested interest in any particular design because no designs have been proposed. We can consider the product objectives in isolation from particular designs. We would like to be able to rank the product objectives from most to least important. It will also be useful for quantitative analysis if we can assign relative weights to the different objectives so that the weights sum to one and the highest weights are associated with the most favored objectives.

Several questions and concerns arise when we consider the problem of weighting and ranking product objectives:

1. Are weightings and rankings of abstract objectives really useful? Can decision making really be reduced to a quantitative basis?
2. Who is doing the ranking? From which perspective (owner, customer, or user) should the ranking be performed?
3. Some comparisons are repugnant to us. Must we rank safety against affordability?
4. Are people truthful when they reveal their preferences? Can we even hope to agree on a ranking of product objectives?

These are all valid and important questions, and none of them can be answered unequivocally. We acknowledge, for example, that decision making cannot be reduced to a purely quantitative basis. Nevertheless, quantitative techniques are designed to ensure a balanced consideration of many factors. Many design teams find these techniques to be useful for discussion. We also recognize that the owner, customer, and user may rank the product objectives in very different ways and that each viewpoint is important. It is also true that some comparisons are repugnant. Safety concerns must be considered in the design process, but it may be unpalatable to compare them in the abstract with other objectives. On the other hand, as humans, we routinely purchase products such as lawn mowers and chain saws that possess a high risk of injury. There is a calculus of risk in our personal decision making; eventually, we will need to address risk in our approach to design. Finally, it is true that people may not be forthcoming in revealing their preferences and that consensus in ranking can be difficult to achieve.

These are all topics worth discussing at length but, for this text, we proceed pragmatically. We assume, for example, that rankings and weightings of product objectives are useful in design and we will show how to use them. We take the perspective of the customer in establishing these rankings. We restrict our definition of safety so that it is comparable to other objectives. Finally, we use a simple breakdown scheme to establish the weightings.

There are four steps in our approach to rank goals and compute relative weights. This approach is known as the *analytic hierarchy process*. It was developed by Professor Thomas L. Saaty in the 1970s and has been used widely in industry and government since then:

1. Group the product objectives into sets and nested subsets of similar objectives.
2. For each group of sets (or subsets), assign fractional weights to each set so that the sum of the weights in the group equals one and the weights are indicative of the relative importance of the set or subset within the group.
3. For each product objective, multiply together all the relative weights of all the nested subsets to which it belongs. Call this quantity the *relative priority* of the product

Table 3.10 Primary and Secondary Product Objectives (Goals)

Make the toy fun for the parent
Make the toy fun for the child
Make the toy appealing to the parent
Make the toy appealing to the child
Make the toy safe for preschoolers
Make the toy easy to put away
Make the toy playable for a long time
Make the toy easy to repair
Make the toy inexpensive relative to similar toys

objective. By construction, the relative priorities of all the product objectives will sum to one.

4. Sort the product objectives by descending order of relative priority. The resulting sort order is the ranking of the product objectives.

Illustration of the Analytic Hierarchy Process Using the Toy Catapult

We illustrate the approach to ranking goals and computing weights using the toy catapult example. Recall that we identified a total of nine goals or product objectives. They are listed in Table 3.10. We have reworded two of these goals to make comparisons between them meaningful. For example, one of the goals was originally stated as "make the toy safe." This is a valid product objective, but if we ask anyone to compare this goal with other goals, it may be an unpalatable comparison.

Instead, for comparison purposes, we limit the scope of the goal to "make the toy safe for preschoolers." In this way, we bring the age-appropriate nature of the toy into the comparison. This makes comparisons of safety with other goals less unpalatable because it is assumed that parents would monitor use of or access to the toy more closely for preschool children. In a similar fashion, we modified the goal originally stated as "make the toy affordable." Again, if the choice is between an affordable toy and an unaffordable one, then comparisons are not meaningful. We capture the relative importance of cost by rewording the goal as "make the toy inexpensive relative to similar toys." Comparisons between this reworded goal and other goals are now likely to be more meaningful.

In considering the safety objective, keep in mind that safety concerns enter the design process in multiple ways. In our requirements analysis, we explicitly created requirements to minimize the risk of the toy to the child, other children, and pets. What we are considering here is how the customer's perception of the toy's safety affects his or her delight in the product. We have also chosen to measure this safety by a metric of the age appropriateness of the toy. As the age-appropriateness of the toy increases (that is, as it becomes less suitable for young children), the parent's satisfaction with the toy may also decrease and the fun or appeal of the toy may not be sufficient to warrant its use. Parents certainly make these judgments on a routine basis so it is reasonable to include safety in this sense as an objective to be weighted and ranked relative to other objectives.

Table 3.11　Product Objectives Grouped by Affinity

Make the Toy Fun		Make the Toy Appealing		Make the Toy Safe for Preschoolers	Other (Easy to Put Away, Durable, Repairable, Inexpensive)			
					Make the Toy Useful (Easy to Put Away, Durable, Repairable)			Make the Toy Inexpensive
Make the toy fun for the parent	Make the toy fun for the child	Make the toy appealing to the parent	Make the toy appealing to the child	Make the toy safe for preschoolers	Make the toy easy to put away	Make the toy play-able for a long time	Make the toy easy to repair	Make the toy inexpensive relative to similar toys

In what follows it will be useful to have *relative priorities* for product objectives with which to compare these objectives. For example, we will later face trade-off decisions such as increasing the launch velocity of the catapult at the expense of increasing the age-appropriate level of the toy. It will be useful in such trade-off situations to have a good sense of the relative importance of the product objectives, such as fun-to-age level.

With such a long list of diverse product objectives as shown in Table 3.10, it is difficult to make relative comparisons. Accordingly, the first step in the analytic hierarchy process is to group these product objectives into nested sets. The goal is to work with many small groupings of objectives for which comparisons are easy and meaningful. We use the affinity process to reorganize the product objectives into the nested groupings shown in Table 3.11.

At the highest level, we have a grouping of four product objective sets: "fun," "appeal," "safety," and "other." The set labeled "other" is divided into two sets: "useful" and "inexpensive." Each of the low-level sets ("fun," "appeal," "safety," "useful," and "inexpensive") is further divided to contain all nine product objectives. To make these groupings clear, we number them in Table 3.12.

Observe that the groupings are nested: Grouping 6 belongs to a single set ("useful") in grouping 2. Grouping 2, in turn, belongs to a single set ("other") in grouping 1. This is called a *product objective hierarchy:* Each product objective can be traced to a single objective at a higher level.

The next step in the analytic hierarchy process is to establish relative weights of importance within each grouping. We consider each grouping, in any order, and assign positive fractions to each of the sets in the grouping. The fractions in each grouping must sum to one and their magnitude must reflect the relative importance of the different product objective sets. In making these judgments, we are taking the point of view of the customer. Ideally, the design team would engage customer representatives, such as customer focus groups, in assessing these weights. Saaty (1980) offers a sophisticated technique, based on pairwise comparisons of objectives, to set these weights. It is outside the scope of this book, where we use a more direct approach.

For example, Table 3.13 shows our personal assessment of these weights. Consider grouping 3 ("fun"). It is reasonable to expect that the parent's primary concern is the fun of the toy from the perspective of the child. Perhaps, the "fun for child" objective should receive twice the weight of "fun for parent." Because the weights for grouping 3 must sum to one, we assign fractions ⅓ to "fun for parent" and ⅔ to "fun for child."

Now consider grouping 4 ("appeal"). In this case, the parent's pleasure in the toy might be more important than it was for the "fun" objective: We use a 40:60 split (0.4 for "appeal to parent" and 0.6 for "appeal to child"). Next, consider grouping 6. From a parent's perspective, we guess that durability would be the most important objective and the other two objectives would be

Table 3.12 Numbered Groupings of Product Objectives

1	Make the toy fun		Make the toy appealing		Make the toy safe for preschoolers	Other (easy to put away, durable, repairable, inexpensive)			
2						Make the toy useful (easy to put away, durable, repairable)			Make the toy inexpensive
3	Make the toy fun for the parent	Make the toy fun for the child							
4			Make the toy appealing to the parent	Make the toy appealing to the child					
5					Make the toy safe for preschoolers				
6						Make the toy easy to put away	Make the toy playable for a long time	Make the toy easy to repair	
7									Make the toy inexpensive relative to similar toys

roughly equal in importance. Accordingly, we assign a weight of 0.4 to "playable for a long time" and 0.3 each to "easy to put away" and "easy to repair." We can ignore groupings 5 and 7 because they are singleton sets (the weight is automatically 1 for each set in these groupings).

Grouping 2 requires us to compare "usefulness" with "inexpensive" on behalf of the parent. "Usefulness" is the name we have given to the aggregate of attributes "easy to put away," "durable," and "easy to repair." How the parent might judge such a trade-off is difficult to ascertain. We have entered a fuzzy area in attempting to make decision-making analytical. Pressing ahead, we assign a relatively low weight to "inexpensive," say 0.2, with the balance, 0.8, going to "usefulness." Personal experience, customer feedback, and design group consensus are all factors to consider at this stage of analysis.

Table 3.13   Relative Weights Assigned within Each Grouping

#	Make the toy fun		Make the toy appealing		Make the toy safe for pre-schoolers	Other (easy to put away, durable, repairable, inexpensive)			
1	Make the toy fun		Make the toy appealing		Make the toy safe for pre-schoolers	Other (easy to put away, durable, repairable, inexpensive)			
	0.25		0.25		0.20	0.30			
2						Make the toy useful (easy to put away, durable, repairable)			Make the toy inexpensive
						0.8			0.2
	Make the toy fun for the parent	Make the toy fun for the child	Make the toy appealing to the parent	Make the toy appealing to the child	Make the toy safe for pre-schoolers	Make the toy easy to put away	Make the toy playable for a long time	Make the toy easy to repair	Make the toy inexpensive relative to similar toys
3	0.333	0.667							
4			0.4	0.6					
5					1				
6						0.3	0.4	0.3	
7									1

Grouping 1 is also difficult, but it is certainly easier to consider four aggregate sets of product objectives ("fun," "appeal," "safety," and "other") than it is to consider the original group of nine product objectives. After some deliberation, we assign weights of 0.25 to "fun," 0.25 to "appeal," 0.20 to "safety," and 0.30 to "other." We gave the lowest weight to safety for two reasons. The first is that to make a toy "fun," there will necessarily be some risk. If the age-appropriate objective ("safety") is given too much importance, we may end up with a very dull toy. The second reason is that all the other objectives are aggregates whose priorities will be subdivided. The "safety" objective is a singleton: It will end with a priority equal to 0.20, whereas the other objectives will naturally end with smaller priorities because of subdivision (see later discussion). We arbitrarily decreased the weight on "safety" and increased the weight on "other" to compensate for the subdivision effect.

The difficult work of assessing weights is done. The remaining steps are mechanical in nature. We first use the relative weights for the groupings to calculate relative priorities for the product objectives. This is done by multiplying together the relative weights of all the groupings in which a product objective is contained. For example, "make the toy easy to repair" is a member of grouping 6. Because of the nested nature of the groupings, it is also contained in groupings 2 and 1. Accordingly, we multiply the weight 0.3 from grouping 6 with weights 0.8 from grouping 2 and 0.30 from grouping 1. After rounding off, the result is 0.07. We refer to this product as the relative priority of product objective "make the toy easy to repair." The calculations for the remaining product objectives are shown in Table 3.14.

The final step in the analytic hierarchy process is to sort the product objectives in decreasing order of relative priority (spreadsheet skill: sort). This order reveals a ranking of the product

Table 3.14 Computing Relative Priorities of Product Objectives

1	Make the toy fun		Make the toy appealing		Make the toy safe for pre-schoolers	Other (easy to put away, durable, repairable, inexpensive)			
	0.25		0.25		0.20	0.30			
2						Make the toy useful (easy to put away, durable, repairable)			Make the toy inexpensive
						0.8			0.2
	Make the toy fun for the parent	Make the toy fun for the child	Make the toy appealing to the parent	Make the toy appealing to the child	Make the toy safe for pre-schoolers	Make the toy easy to put away	Make the toy playable for a long time	Make the toy easy to repair	Make the toy inexpensive relative to similar toys
3	0.333	0.667							
4			0.4	0.6					
5					1				
6						0.3	0.4	0.3	
7									1
Formula	= 0.333 * 0.25	= 0.667 * 0.25	= 0.4 * 0.25	= 0.6 * 0.25	= 1 * 0.2	= 0.3 * 0.8 * 0.3	= 0.4 * 0.8 * 0.3	= 0.3 * 0.8 * 0.3	= 1 * 0.2 * 0.3
Rounded result	0.08	0.17	0.10	0.15	0.20	0.07	0.10	0.07	0.06

objectives that may be useful. Table 3.15 shows the result of transposing the rounded results from Table 3.14 and sorting them in decreasing order of relative priority. Not surprisingly, the "safety" product objective surfaces as the highest ranked product objective.

By construction, the relative priorities sum to one, subject to round-off error. Figure 3.2 shows the relative priorities as a pie chart.

We have not abandoned our mission statement for the design of a toy catapult. It is still our mission to "delight both parent and child." Table 3.15 is very much secondary to our mission. Nevertheless, by ranking and weighting the customer priorities, we gain valuable insight into what it will mean to create delight.

Benchmark Competition on Measures of Effectiveness

A *benchmark* is a performance test of hardware or software. In this section, we select two commercially available catapult toys and conduct a mock benchmarking exercise to provide data

Table 3.15 Relative Priorities and Ranking of Product Objectives

Product Objective	Relative Priority	Rank
Make the toy safe for preschoolers	0.20	1
Make the toy fun for the child	0.17	2
Make the toy appealing to the child	0.15	3
Make the toy appealing to the parent	0.10	4
Make the toy playable for a long time	0.10	5
Make the toy fun for the parent	0.08	6
Make the toy easy to put away	0.07	7
Make the toy easy to repair	0.07	8
Make the toy inexpensive relative to similar toys	0.06	9
Total	1.00	

for the design of our own toy catapult. Numerous toy catapults are available for sale on the Internet. Figure 3.3 and Figure 3.4 illustrate two of the many choices: the Catapult Kit by Pathfinders and the Scheich Catapult Replica, respectively. We have purchased and evaluated both of them.

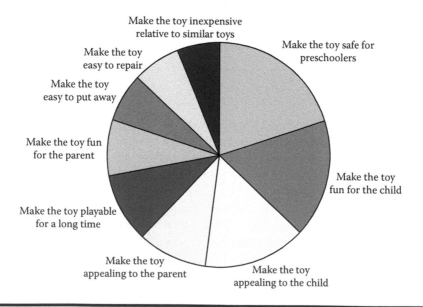

Figure 3.2 Relative priorities of product objectives.

Figure 3.3   Catapult Kit by Pathfinders.

In Table 3.16 we collect data on the two catapults. These data are taken from the Web sites for the two products together with personal estimates. We tested the throwing capability of each toy by launching a small eraser. The eraser was too heavy for the replica to throw. The catapult built from a kit threw it across the room. One online reviewer of the catapult kit complained that the top crossbar broke after only 2 hours of play.

Figure 3.4   Catapult Replica by Schleich.

Table 3.16 Benchmarking Data for Commercially Available Toy Catapults

	Catapult Kit	*Catapult Replica*
Price	$28.95	$19.95
Recommended ages	10+	4+
Throwable distance (eraser)	15 ft	0 ft
Material	Wood	Hand-painted vinyl
Assembly time	2 Hours	2 Minutes
Time to failure	2 Hours (according to one reviewer)	Indefinite
Dimensions	11 × 5 × 5	7.5 × 4.5 × 4.5

Based on these data, we proceed to rank the two commercially available toys in each of the customer product attributes on a scale of 1–5, where 1 represents "poor" and 5 represents "excellent." This is a subjective judgment. Ideally, this step is done by a customer focus group. Table 3.17 presents our judgments.

As we can see from the customer perceptions, these are two very different products. There are significant differences between their rankings in every category. No one toy dominates in every product objective. There is certainly an opportunity to improve on these scores for our target market.

Graphical Representation of Benchmarking Data

Before leaving the discussion on benchmarking, we note that tabular presentations such as Table 3.17 are only one way to present multidimensional data for comparison purposes. Two graphical techniques are a slider bar representation (Figure 3.5) and a radar chart (Figure 3.6). Radar charts are one of the chart types supported by MS Excel.

Table 3.17 Customer Perceptions of Competitors' Products

	Catapult Kit	*Catapult Replica*
Visual appeal: 1 (unattractive)–5 (very attractive)	2	5
Performance: 1 (poor)–5 (excellent)	5	1
Safety: 1 (age 12+)–5 (age 3+)	1.5	4.5
Reliability: 1 (poor)–5 (excellent)	1.5	5
Ease of repair: 1 (poor)–5 (excellent)	4.5	2.5
Ease of storage: 1 (poor)–5 (excellent)	3.5	4
Affordability: 1 (poor)–5 (excellent)	3	4

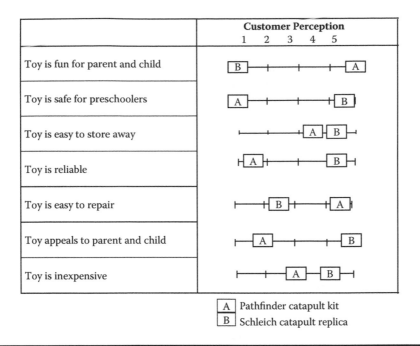

Figure 3.5   Slider bar representation of benchmarking data.

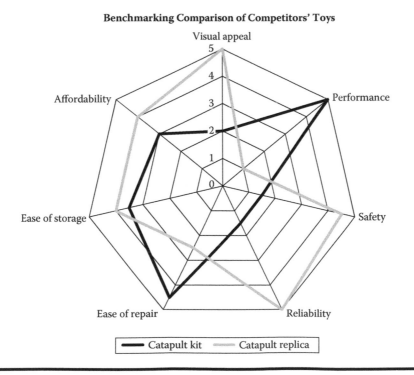

Figure 3.6   Radar chart representation of benchmarking data.

Translate to Technical Requirements

We find ourselves now at the interface between design and engineering. We have identified product objectives that are important to the customer and we have devised ways to measure our success and the success of our competitors against those objectives. We have even developed a sense, through the analytic hierarchy process, of the relative importance of those product objectives. The next step takes us into the engineering world. We now want to map these customer-focused product objectives to specific design and engineering characteristics of the product and set target performance measures for these characteristics. That is, we want to try to convert the design problem into an engineering problem.

This transition between a customer focus and an engineering focus can be difficult. People tend to excel in one area but not the other. Designers, marketers, and other customer-focused people in an organization often speak what seems like a different language from that of the engineers. Engineers are capable of solving complex technical problems but they need to know what problems to solve. They need the guidance of the design organization. Failure to communicate across this interface between design and engineering is a frequent source of frustration and, sometimes, humor (Figure 3.7).

The focus of this section is on how to make the transition from customer objectives to detailed technical performance measures.

Requirements for a Brake Handle

At an alumni event, I was talking with a former student from 25 years ago. He is now an industrial manager, currently based in China and managing the quality assurance

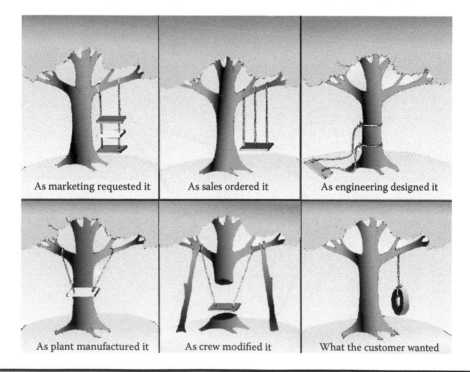

Figure 3.7   A classic cartoon on the failure to communicate. (Original author unknown. This version courtesy of Merrisa Carter.)

function for a leather products manufacturing company. I mentioned to him that many people from industry had impressed upon me the importance of establishing clear requirements early in the design process. He became quite animated and added his story to the list. He had just recently finished writing down requirements for a leather-covered emergency brake handle to be produced by his company for a German automaker. There were 120 requirements for this simple product. Until he had written those requirements down, the project had been extremely frustrating.

My former student's company had not been dealing directly with the German automaker. Instead, the order had come through another company headed by a man who was a great salesman but was rather vague on technical details. Months had gone by, during which time the company had tried to design the product and develop manufacturing plans. The company was at considerable risk both from the perspective of financial investment in manufacturing equipment and also from a product liability standpoint. Finally, my former student insisted that the project would not proceed until the sales company had provided a firm set of requirements. Starting from only four basic objectives, which included safety and aesthetics, he negotiated the precise set of requirements—120 of them—that defined the product. Since then, his company has been able to pursue the project with confidence.

A Multimillion Pound Lawsuit

British broadcaster Sky filed suit in 2007 against the software development firm EDS, claiming the vendor was fraudulent. It alleged that EDS had won the contract by exaggerating its abilities and resources and was then unable to deliver on its promises. Sky claimed £709 million in damages. In its defense, EDS said that the problem was not their ability to create the software, but Sky's inability to define what the requirements were.

Representing EDS, barrister Mark Barnes [stated]"…three years into the contract,… the project specifications remained so unclear that a special team [had to be set up] in order to define the exact requirements of the project."

Leo King (2007)

The House of Quality

One of the best tools for communication between design and engineering is the so-called "house of quality" technique developed by Professors Shigeru Mizuno and Yoji Akao (1994) in Japan in the late 1960s. It is part of a comprehensive approach to quality design systems known as *quality function deployment* (QFD). Even if you are not an engineer, it is valuable to understand this technique so that you can participate in discussions that use it.

The *house of quality* is a matrix approach for defining the relationship between customer desires and product characteristics (Hauser and Clausing, 1988). On a single page, it brings together all of the information that is relevant for a discussion between designers and engineers. It does so by linking together matrices that summarize important relationships, using the metaphor of a house to put them together. Figure 3.8 is a depiction of the house with the following different matrices forming the different parts of the building:

1. front porch: the customer product objectives and their relative importance;
2. back porch: the customer perceptions of competing products;

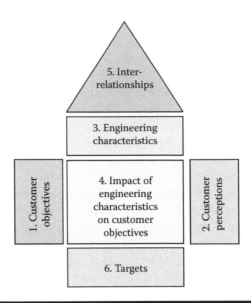

Figure 3.8   Linked matrices forming the "house of quality."

3. second floor: the primary engineering characteristics of the product;
4. main floor: the impact that the different engineering characteristics have upon the customer product objectives;
5. roof or attic: the interrelationships of the engineering characteristics; and
6. basement: the target performance measures established for the different engineering characteristics.

Numerous software packages and spreadsheet add-ins exist to create a house of quality. A template for MS Excel can be downloaded from the Getting Design Right Web site. In creating this template, we found it difficult to create the roof portion of the graphic. Also, in teaching this technique, we have found that students think of the roof of the house as a more difficult and isolated concept than the other matrices simply because of its shape and orientation. Our template, therefore, treats the roof as shown in Figure 3.9. It is a simpler orientation and it more clearly reveals that there is nothing special about the matrix of engineering interrelationships. It is simply the lower triangular portion of a symmetric matrix. We have nicknamed this the "shed of quality" because its visual appearance is so different from the more traditional house of quality. Both house and shed convey the same information. We simply find the shed to be easier to work with.

The goal of the house of quality technique is to establish targets for technical performance measures that are consistent with customer product objectives. In the house metaphor, these targets form the basement of the house. All the information that is relevant to making decisions about these targets is found in the other rooms of the house. The technique is simply a way to bring all of this information together for a decision. It does not actually specify how you make the decision or set the targets.

The process of building a house of quality is as follows:

1. Identify and bundle customer focused product objectives.
2. Rank and weight the product objectives from the customer's perspective.

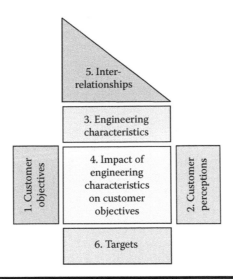

Figure 3.9 The "shed of quality," an easier graphic to create.

3. Benchmark the competition on the product objectives.
4. Identify the engineering characteristics that are most relevant to product objectives.
5. Assess the impact (positive or negative) that the engineering characteristics have on the product objectives.
6. Relate the engineering characteristics to each other.
7. Identify units of measure, cost, difficulty, and other aspects of the engineering characteristics.
8. Benchmark the competitors on the engineering characteristics.
9. Decide on target technical performance measures for each engineering characteristic.

The first three steps (identifying product objectives, ranking and weighting them, and benchmarking competitive products against them) have been covered in earlier sections of this chapter. In this section, we focus on the remaining six steps.

Example: A House of Quality for the Toy Catapult

Figure 3.10 displays a house of quality for the toy catapult. Observe that the list of customer attributes and their weights of relative importance comes from Table 3.15. The slider bar representation of customer perceptions comes from Figure 3.5. Thus, the front and back porch of our house come from our earlier customer analysis and product benchmarking. We examine the remaining rooms of the house in detail in the next subsections.

Identify Engineering Characteristics

The second floor of the house of quality is the list of *engineering characteristics* together with their direction of change. Figure 3.11 shows the contents of the second floor. The process of identifying engineering characteristics is to think of measurable product attributes that would strongly affect the customer attributes. For example, the fun of the toy is related to its ability to throw projectiles. Presumably, larger projectiles and an ability to throw them further will make the product more

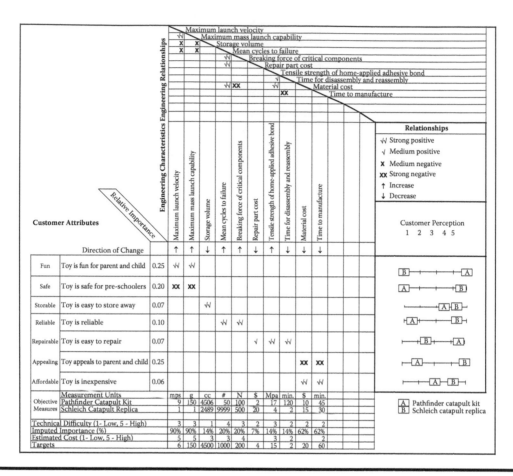

Figure 3.10 House of quality for the toy catapult.

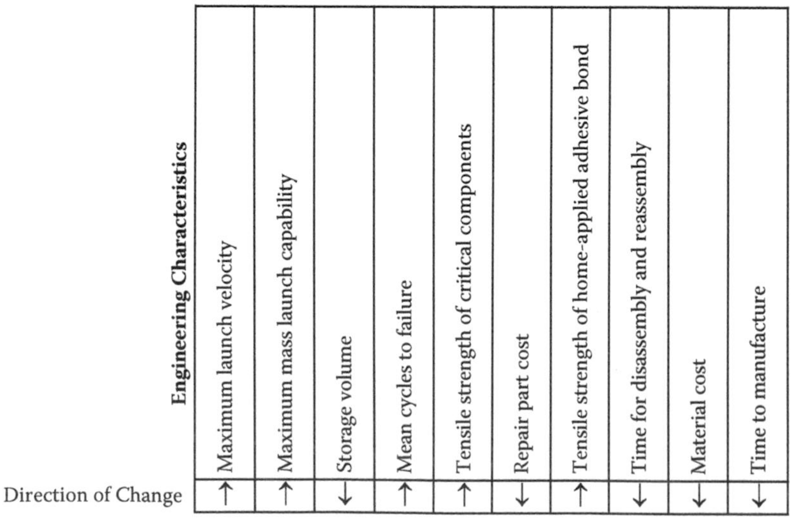

Figure 3.11 The second floor: engineering characteristics and direction of change.

desirable from a fun perspective. How do we measure the size of the projectiles and the distance they can be thrown? There are several possibilities. We have chosen to capture those aspects by measuring "maximum launch velocity" and "maximum mass launch capacity." In like manner, we work through the other customer attributes and identify relevant engineering characteristics.

As another example, consider the product attribute of "reliability." We note that the most likely form of product failure is for the force of the catapult to weaken and break some structural component, such as the crossbar. (A customer complaint for the Pathfinders kit mentioned that the crossbar broke after 2 hours of use.) Not all designs will use a crossbar, so we list the engineering characteristics in a more generic fashion: "mean cycles to failure" and "tensile strength of critical components." As a final example, when considering repairability, we noted that many toys cannot be repaired at home because household glues are ineffective on some materials. This was captured under the engineering characteristic "tensile strength of home-applied adhesive bond."

Clearly, this is an area where engineering knowledge becomes important. Understand, however, that we are at the threshold between design and engineering. It is important for the customer's voice to be heard in this process. For example, if customers say they want a "peppy" car (customer attribute), that doesn't mean, necessarily, that they want a vehicle with high horsepower (engineering characteristic). It may mean that they want high torque at low speeds (engineering characteristic). The designers and engineers must work together to see that customer desires are correctly translated into product terms.

Note that we list a direction of change under each engineering characteristic. All stated relationships with this engineering characteristic will be in terms of this direction of change. That way, we can form sentences from the data of the house of quality, such as "when we *increase* maximum launch velocity, we expect that the toy will be *more* fun for parent and child."

Map Engineering Characteristics to Customer Attributes

The main floor of the house of quality is the matrix formed from the customer attributes as rows and the engineering characteristics as columns. It is here that you can tell the story (or read the story) of how the engineering characteristics affect the customer attributes. Figure 3.12 shows the main floor (with the front porch) for the toy catapult example. Figure 3.13 displays a typical legend of symbols used to describe relationships. Using this legend, we can read the following statements from Figure 3.12:

- Increasing maximum launch velocity strongly affects fun in a positive direction (there is a double check in the "fun" row and "launch velocity" column).
- Increasing maximum launch velocity strongly affects child safety in a negative direction (a double x in the "safety" row and "launch velocity" column).
- Decreasing the cost of repair parts moderately affects repairability in a positive direction.
- Decreasing the material cost of the toy (using cheaper materials) strongly affects the appeal of the toy in a negative direction.

If an engineering characteristic has little or no impact on any customer attribute (there are no checks or x's in the column), then it does not belong in the house of quality at all. It may be relevant at a lower level of design. For example, the color of glue used in construction may become relevant when we get to a deeper level of design, but it is not of major concern at this stage. The house of quality is focused on high-level system characteristics.

Observe how the main floor of the house of quality captures important trade-offs that must be faced by the design team. Increasing launch velocity improves the fun value of the toy but

Customer Attributes		Relative Importance	Maximum launch velocity	Maximum mass launch capability	Storage volume	Mean cycles to failure	Breaking force of critical components	Repair part cost	Tensile strength of home-applied adhesive bond	Time for disassembly and reassembly	Material cost	Time to manufacture
Direction of Change			↑	↑	↓	↑	↑	↓	↑	↓	↓	↓
Fun	Toy is fun for parent and child	0.25	√√	√√								
Safe	Toy is safe for preschoolers	0.20	XX	XX								
Stor-able	Toy is easy to store away	0.07			√√							
Reli-able	Toy is reliable	0.10				√√	√√					
Repair-able	Toy is easy to repair	0.07						√	√√	√√		
Appeal-ing	Toy appeals to parent and child	0.25									XX	XX
Afford-able	Toy is inexpensive	0.06									√√	√√

Figure 3.12 Main floor: map of engineering characteristics to customer attributes.

reduces safety. Reducing material cost (using cheaper materials) makes the toy more affordable but can lessen the appeal of the toy. At the same time, we can see the relative importance of the customer attributes and use that to guide our thinking. For example, we should not be too eager to reduce material cost because affordability is not as important as the appeal of the toy. All of these relationships are captured in an easily readable form for everyone on the team to see and discuss.

Relationships	
√√	Strong positive
√	Medium positive
X	Medium negative
XX	Strong negative
↑	Increase
↓	Decrease

Figure 3.13 Legend of relationships.

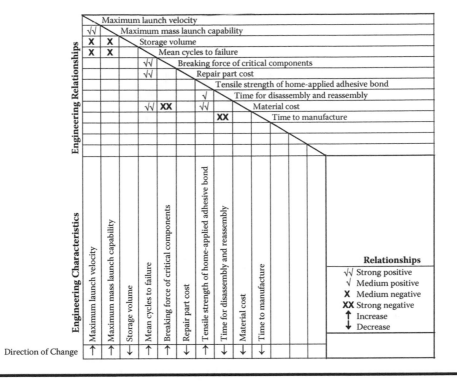

Figure 3.14   The attic: engineering interrelationships.

Document Engineering Interrelationships

Besides the relationships between engineering characteristics and customer attributes, other important trade-offs need to be understood before setting firm targets. Among these are trade-offs that occur between engineering characteristics themselves. The purpose of the roof or attic of the house is to capture these relationships. Figure 3.14 displays the attic of the house of quality. It is the lower triangular portion of a square matrix in which the list of engineering characteristics forms both the rows and columns (in the same order). It is not necessary to show the upper triangular portion because the matrix is understood to be symmetric. From this figure, we can read the following relationships:

- Increasing the maximum launch velocity is strongly positively related to increasing the maximum mass launch capability (there are two checks in the row "maximum mass launch capability" and column "maximum launch velocity").
- Increasing maximum launch velocity or maximum mass launch capability is moderately negatively related to decreasing storage volume (there is a single x in the row "storage volume" for these characteristics); that is, to make a more powerful catapult, it will likely be bigger.
- Increasing the tensile strength of critical components is strongly negatively related to decreasing material cost (i.e., increasing tensile strength requires more expensive materials).

Thus, the attic is the place for engineers to tell their side of the design story: why it is difficult to improve one product attribute without making other attributes worse.

Engineering Characteristics	Maximum launch velocity	Maximum mass launch capability	Storage volume	Mean cycles to failure	Breaking force of critical components	Repair part cost	Tensile strength of home-applied adhesive bond	Time for disassembly and reassembly	Material cost	Time to manufacture			
Direction of Change	↑	↑	↓	↑	↑	↓	↑	↓	↓	↓			
Measurement units	mps	g	cc	#	N	$	Mpa	min.	$	min.			
Pathfinder catapult kit	9	150	4,506	50	100	2	17	120	10	45			
Schleich catapult replica	1	1	2,489	9,999	500	20	4	2	15	30			

(Objective Measures)

Figure 3.15 Units of measure and engineering benchmarks.

As you can see by now, the front and back porches, the second and main floors, and the attic capture an interesting set of perspectives on the design problem. You can imagine how involved design discussions can get and why it might be useful to have a map like the house of quality to keep track of everything.

Identify Units of Measure and Benchmark Competitors

In the basement of the house, we focus on the engineering characteristics and develop more detailed measurement schemes. Figure 3.15 shows a portion of the basement. We start by deciding on units of measure for the different characteristics. For the toy catapult, we have chosen to use the metric system of measurement: meters per second (mps) for velocity, grams (g) for mass, cubic centimeters (cc) for volume, newtons (N) for force, and megapascals (Mpa) for tensile strength. We purchased the two competing products and conducted simple tests to estimate some of the parameters such as maximum launch velocity and maximum launch capacity. The tests were very informal. They are for illustration purposes only and are not intended for serious use. For other parameters, we used estimates based on the design and materials as well as customer feedback (posted on Web sites) on reliability.

As you can see from Figure 3.15, no one product dominates in all categories. The Schleich catapult is made from durable materials and has a sturdy construction, but it had almost a negligible ability to throw a projectile. The Pathfinder has an excellent throwing capability but uses lightweight materials. It is not hard to see why the Pathfinder crossbar might break easily: It has a hole drilled in the center that weakens the structure considerably.

Observe one benefit of the technical benchmarking study: We now have reasonable ranges for the engineering characteristics. We can see that to have a product that is as fun for parent and child as the Pathfinder (see the customer perceptions in the back porch), we will need to have a launch velocity approaching 9 mps and a maximum mass launch capacity approaching 150 g. However, to have a product as reliable as the Schleich, we will need the breaking force of critical components to be closer to 500 N than to the 100 N estimated for the Pathfinder.

Determine Target Technical Performance Measures

Before we estimate the target technical performance measures, we note that many design and engineering teams find it useful to supplement the basement matrix with three more rows (technical difficulty, imputed importance, and estimated cost) as shown in Figure 3.16 for the toy catapult. The discussion in setting targets will inevitably involve discussions around these attributes, so it is good to include them in the matrix.

Technical difficulty is a subjective rating by the engineers of the challenges to make changes (in the direction of change indicated) in the engineering characteristics. For example, we rated the technical difficulty of reducing storage volume to be quite low because that is simply a matter of scaling the design. We felt that the greatest technical challenge was to design a catapult to be fast, strong, and reliable. If the catapult has high launch velocity and high mass launch capability, then the experience of the Pathfinder design shows that it is not easy to make it reliable as well.

We also made subjective ratings of the cost of making changes in the engineering characteristics (except for the cost characteristics). These are correlated with the technical difficulty because we believed that the attributes of fast, strong, and reliable would drive the cost of the product.

The final attribute considered, "imputed importance," is interesting. What we want to have in this row is a measure of how important this engineering characteristic is to the customer. We have no direct way of measuring how the customer values something like "tensile strength of home-applied adhesive bond," so we must deduce or impute a value. One simple way of establishing that measure is to base it on the main floor of the house of quality, the matrix that relates engineering characteristics to customer attributes. Observe in Figure 3.12, for example, that the importance of the characteristic "tensile strength of home-applied adhesive bond" comes from the fact that it has two check marks ("is strongly positively related to") in the row for "easy to repair." Now, our customer surveys revealed that the relative importance of the "easy to repair" attribute is 7% (see the column "relative importance" in Figure 3.12). Multiplying 7% by 2 (for the two check marks) yields 14%. We take this number in Figure 3.16 to represent the relative importance of "tensile strength of home-applied adhesive bond."

How shall we measure the relative importance of an attribute such as "maximum launch velocity"? Observe in Figure 3.12 that this engineering characteristic affects "fun for parent and child" in a positive manner (two checks) and "safe for child" in a negative manner (two x's). It would not make sense to subtract the x's from the checks because that would indicate that launch velocity is unimportant. Instead, we add the negative impact to the positive impact and compute the imputed importance as twice the relative importance of "fun for parent and child" plus twice the relative importance of "safe for child," resulting in a total imputed importance of 90%.

The imputed importance row is useful in guiding the team to focus its attention on engineering characteristics that will have the greatest impact on customer delight. For example, Figure 3.16 leads us to focus attention on maximum launch velocity and maximum mass launch capacity (90% imputed importance) and leads us away from worrying too much about repair part cost (7% imputed importance).

We are now in a position to estimate the *target technical performance measures*. The house of quality approach does not specify a way to set these targets. The main value of the house of quality is in pulling together all of the information that is valuable in setting these targets. Look back at the big picture of Figure 3.10. Here, on a single page, is a map to all the relationships that typically arise or that should arise in a target-setting team meeting. One of the problems in team meetings is the

		Maximum launch velocity	Maximum mass launch capability	Storage volume	Mean cycles to failure	Breaking force of critical components	Repair part cost	Tensile strength of home-applied adhesive bond	Time for disassembly and reassembly	Material cost	Time to manufacture
	Direction of Change	↑	↑	↓	↑	↑	↓	↑	↓	↓	↓
Objective Measures	Measurement units	mps	g	cc	#	N	$	Mpa	min.	$	min.
	Pathfinder catapult kit	9	150	4,506	50	100	2	17	120	10	45
	Schleich catapult replica	1	1	2,489	9,999	500	20	4	2	15	30
Technical difficulty (1-Low, 5-High)		3	3	1	4	3	2	3	2	2	2
Imputed importance (%)		90%	90%	14%	20%	20%	7%	14%	14%	62%	62%
Estimated cost (1-Low, 5-High)		5	5	3	3	4		3	2		2
Targets		6	150	4,500	1,000	200	4	15	2	20	60

Figure 3.16 Setting target technical performance measures.

tendency for discussion to cycle endlessly. Multidimensional linked matrices, such as the house of quality, are a useful technique to make sure everyone's perspective is captured. People tend to relax when they see that their particular perspective has been considered by the rest of the team.

You can see the targets we set for the toy catapult in Figure 3.16. Because of safety concerns, we decided to make the toy less powerful than the Pathfinder toy. We believed, however, that it was possible to make it much more reliable. We also decided that storage volume, time to manufacture, and material cost could be sacrificed to achieve our other goals, so we set generous upper limits in those engineering characteristics.

The objective of the house of quality exercise is to set specific requirements for technical performance measures. It remains to write these requirements out in a formal, unambiguous fashion. That is not hard to do, as shown in Table 3.18, given all the work that has gone into the house of quality. Observe that we have called these "originating requirements" because they cannot be derived from any previously known system requirement. We have continued the numbering scheme from Table 2.19, so our first technical performance requirement is OR.14.

Collect and Rationalize System-Level Requirements

Using behavior analysis, we arrived at a set of originating requirements for functionality, Table 2.19. Using customer analysis and the house of quality approach, we arrived at a set of originating requirements for technical performance measures, Table 3.18. We combine all of these requirements into a list of originating system requirements in Table 3.19.

Before proceeding any further with design, we should check our list of originating requirements for consistency and clarity. We observe immediately, for example, that requirement OR.11 was a vague statement of reliability that came out of our behavior analysis. Requirement OR.17 is a much more specific

Table 3.18 Target Technical Performance Measures

Index	Originating Requirements	Abstract Name
OR.14	The maximum launch velocity of the toy catapult shall be at least 6 mps	Maximum launch velocity
OR.15	The maximum mass launch capability of the toy catapult shall be at least 150 g	Maximum mass launch capability
OR.16	The storage volume of the toy catapult shall not exceed 4,500 cc	Storage volume
OR.17	The mean cycles to failure of the toy catapult shall be at least 1,000	Mean cycles to failure
OR.18	The breaking force of a critical component such as the crossbar shall be at least 200 N	Breaking force of critical components
OR.19	The cost of a replacement for a critical component such as a crossbar shall not exceed $4	Repair part cost
OR.20	If a critical component such as a crossbar breaks and it is repaired using typical home-applied adhesive, the tensile strength of the home-applied adhesive bond shall be at least 15 MPa	Tensile strength of home-applied adhesive bond
OR.21	The time required for disassembly and reassembly shall not exceed 120 minutes	Time for disassembly and reassembly
OR.22	The material cost of the toy catapult shall not exceed $20	Material cost
OR.23	The time to manufacture the toy catapult shall not exceed 60 minutes per toy	Time to manufacture

statement of reliability that came from the house of quality analysis. We do not need both requirements, and OR.17 is clearly superior. Consequently, we have stricken OR.11 from further consideration.

What is different about this approach from the way you currently conduct design? It takes restraint not to charge ahead with a design idea as soon as it enters your head. Look again at everything we have done before considering the design problem itself. We have studied the customer and the context. We have looked at primary behaviors as well as secondary behaviors. We have developed measures of effectiveness to determine whether the product or service is a success. We have identified attributes of the design that are important to the customers and users. We have benchmarked the competition against these attributes and we have developed target performance measures that our design must achieve if it is to be competitive. We have summarized all of this analysis in a set of originating requirements that define not only what the proposed product or service must do, but also how well it must do it and under what constraints we must develop it. Experienced designers will emphasize to you how important it is not to engage in design until you have a clear set of requirements to work from. You are now in command of a set of tools and processes to capture those requirements.

Table 3.19 Collected System-Level Requirements: Functional and Technical Performance Measures

Index	Originating Requirements	Abstract Name
OR.1	The system shall detect the command to arm.	Detect
OR.2	The system shall secure its receptacle in the armed position. The force required to arm the system may be supplied by the system (e.g., battery power) or by the child.	Arm
OR.3	The system shall hold the projectile in its receptacle.	Load
OR.4	The system shall detect the command to release from a passing toy (lateral direction) and also from a child or falling projectile (vertical direction).	Trigger
OR.5	The system shall be capable of ejecting lightweight balls and dolls from its receptacle.	Eject
OR.6	The system shall notify the parent of the dangers and inherent risks of the system.	Warn
OR.7	The system shall suggest to the parent the appropriate age of the child to use the system.	Suggest age
OR.8	The moving parts of the system shall be prevented from striking the face of a child with sufficient force to cause a bruise.	Protect face
OR.9	The system shall not launch projectiles from the child's domain with sufficient force to cause damage to the child or the projectile.	Constrain launch force
OR.10	The system shall enable the escape of a pet rodent from the receptacle.	Enable pet escape
OR.11	~~The system shall successfully complete each cycle of use for many cycles.~~	~~Repeat~~
OR.12	The system shall survive a collision with a hard surface and remain in working order.	Survive collision
OR.13	Verification of all requirements shall be conducted in such a way as to preserve the health of human and animal subjects.	Test safely
OR.14	The maximum launch velocity of the toy catapult shall be at least 6 mps.	Maximum launch velocity
OR.15	The maximum mass launch capability of the toy catapult shall be at least 150 g.	Maximum mass launch capability
OR.16	The storage volume of the toy catapult shall not exceed 4,500 cc.	Storage volume
OR.17	The mean cycles to failure of the toy catapult shall be at least 1,000.	Mean cycles to failure

Table 3.19 Collected System-Level Requirements: Functional and Technical Performance Measures (Continued)

Index	Originating Requirements	Abstract Name
OR.18	The breaking force of a critical component such as the crossbar shall be at least 200 N.	Breaking force of critical components
OR.19	The cost of a replacement for a critical component such as a crossbar shall not exceed $4.	Repair part cost
OR.20	If a critical component such as a crossbar breaks and it is repaired using typical home-applied adhesive, the tensile strength of home-applied adhesive bond shall be at least 15 MPa.	Tensile strength of home-applied adhesive bond
OR.21	The time required for disassembly and reassembly shall not exceed 120 minutes.	Time for disassembly and reassembly
OR.22	The material cost of the toy catapult shall not exceed $20.	Material cost
OR.23	The time to manufacture the toy catapult shall not exceed 60 minutes per toy.	Time to manufacture

Identify the Customer Value Proposition

By this point in the process, we have studied both the customer and the competition and we have defined both functional requirements and technical performance measures that will define our system and, we hope, distinguish it from the competition. This is a good point in the process to articulate a customer value proposition. Recall that a mission statement is a statement of the goals of the project from the perspective of the owner. A value proposition looks at the project from the perspective of a customer.

A *customer value proposition* is a statement, from the perspective of the prospective customer, of why the proposed system should be purchased rather than any of its competing alternatives. It is important to consider the alternatives. The value proposition is not simply a list of the features of the system that will appeal to a customer. Instead, it seeks to identify a compelling reason why, for a particular type of customer, the proposed system offers greater value (net of cost) than any of the alternatives.

This is a critical juncture. If your value proposition is weak—that is, if you cannot convincingly argue why a customer would choose your product over other products in the marketplace—then the risk of market failure is high. It might be better to cancel the project than to waste the time and money required to take it to market. Alternatively, you may backtrack and revise your product concept, functional requirements, or technical performance requirements until you can present a strong value proposition.

Guidelines for Writing a Customer Value Proposition

Jill Konrath (2002) provides a few examples of weak customer value propositions:

- It's the most technologically advanced and robust system on the market.
- We improve communication and morale.
- We offer training classes in a wide variety of areas.
- Our product was rated the best-in-class by leading authorities.

As Konrath explains, the natural response to each of these propositions is, "So what?" Only by continually asking this "so what?" question will you uncover what it is about your product or service that is of real interest and value to your potential customers.

A strong value proposition is one that connects with the customers' goals. For an industrial client, your product or service has value if it reduces the client's costs, increases revenues, speeds up the product development process, reduces uncertainties, or has some other such tangible benefit.

In the case of a consumer product or service, there are other dimensions, such as aesthetics, image, and lifestyle to consider. But in each case, value is what is perceived by the customer, rather than by the designer.

Example: Toy Catapult Customer Value Proposition

To illustrate the idea of a customer value proposition, we restrict consideration of customer alternatives to the two competing products identified in the benchmarking study, the Catapult Kit by Pathfinders and the Scheich Catapult Replica.

Comparing our design concept with the Catapult Kit, we argue that our concept will provide more fun for the parent and child because the trigger mechanism will be more versatile (it can be triggered from both the vertical and horizontal directions) and it will be more reliable. Note that one of these advantages (trigger versatility) came out of our functional requirements and the other (reliability) came out of our technical performance requirements. In other respects (fun for child, appeal, etc.), our concept will score quite similarly with our target customer as the Catapult Kit does. However, we note that it would also be possible to increase the appeal for young children of our design concept dramatically over the Catapult Kit by giving the toy a personality.

Comparing our design concept with the Catapult Replica, we argue that, like the Catapult Kit, our design concept will be much more fun for both parent and child.

In summary, the customer value proposition is that a toy catapult of our design will be more fun for both parent and child than the Catapult Replica or the Catapult Kit because of its performance and versatility, and it will be more reliable than the Catapult Kit. Furthermore, giving the toy a personality will give it stronger appeal to the child than either the Catapult Kit or the Catapult Replica.

Is this customer value proposition strong enough to justify continuing with the project? This is a hobbyist project and financial profit is not a concern, so the answer is "yes." However, even if it were a commercial project, it would appear that we have identified a potential market niche that might be worth pursuing.

Summary

In this chapter, we have moved from general goals, stated in the vernacular language of the customer ("make it fun") to specific targets for engineering product characteristics ("the maximum launch velocity of the toy catapult shall be at least 6 mps"). This is a valuable skill to acquire because so few people are really expert at it. It is a skill that is right at the boundary between customer-focused design and product-focused engineering. Our example has focused on a mechanical toy, but the same skill is needed in designing consumer electronics, applications software, Web services, home and office building architecture, amusement parks, military aircraft, and much more.

We used the goal–question–metric approach to move from general customer goals to specific ways to measure achievement of those goals. The resulting metrics became our measures of effectiveness for the project and enabled us to benchmark different products against each other with respect to the different goals. It also will enable us to measure the success of our design against those goals. We then used the analytic hierarchy process to estimate relative priorities from the customer's perspective to attach to the product goals, also called product objectives. This process enabled us to work with small, manageable subsets and aggregations of product objectives. It was easier to assign relative weights of importance when working with subsets of objectives than working with the original long list of objectives. We demonstrated three different ways to present multidimensional benchmarking data: tables, slider bars, and radar charts.

Using the house of quality approach, we systematically assembled all the information that would be useful for discussion in a design and engineering meeting in order to establish target engineering product characteristics. We used multidimensional linked matrices (i.e., the house of quality) to display all of the relevant information on a single page. We did not provide a technique to set the targets. The targets emerge as the consensus result of team discussion of the many issues involved. The targets can be rewritten easily as originating technical performance requirements and integrated with the list of functional requirements derived from behavioral analysis.

Finally, having looked at both customers and competitors, we formulated a customer value proposition and decided whether it was strong enough to continue with the project.

Discussions

1. Debate the adage *caveat emptor* ("let the buyer beware") in the case of the Relaxacizor product ban.
2. Pick an unusual invention, such as the Segway™ Human Transporter, pick a potential market (who would buy it?), and propose a customer value proposition aimed at that market. How successful is the invention in that market at the current time?

Exercises

1. Using at least four of the customer categories ("voice of the customer") that you developed in Exercise 2.7, apply the goal–question–metric approach to develop measures of effectiveness and data collection ideas for the BathBot. Present your results as in Table 3.6. If you prefer to continue with your own product idea, rather than use the BathBot case, start with at least four high-level customer goals that you think might come out of a customer comment analysis.
2. Refer to the modified version of the BathBot house of quality (Figure 3.17). Explain your answers to the following questions with reference to entries in the house of quality (not to your own domain expertise):
   a. Why might "actuator speed" have a greater imputed importance than "arm positional precision"?
   b. From a competitive perspective, what is the range of actuator speeds that should be considered?
   c. Describe the trade-offs involved in selecting a target for actuator speed. That is, explain how the design would benefit and how the design would suffer (from the customer's

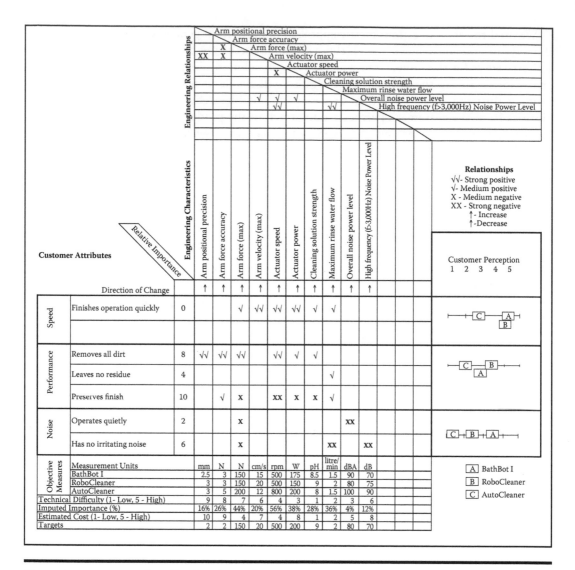

Figure 3.17 House of quality for the BathBot.

perspective) if the target were increased. (Hint: Some of the trade-offs trace through the engineering design relationship matrix.)

d. The designers chose to set high values both for arm positional accuracy (measured as a low tolerance for error) and for arm velocity and a relatively low value for actuator speed. In view of the trade-offs involved in increasing actuator speed, use the entries in the customer design relationship matrix to explain why this might be a good strategy.

e. What can you say about the technical difficulty and cost of this strategy that the designers have chosen? That is, is it easier and less expensive or more difficult and more expensive than working to increase actuator speed and dealing with the resulting trade-offs? State the assumptions you made to answer this question.

3. Create a house of quality for a design problem of your choice.

References

Basili, V. R., and D. Weiss. 1984. A methodology for collecting valid software engineering data. *IEEE Transactions on Software Engineering* November: 728–738.

Caswell, C. 2008. Herreshoff heritage: Herreshoff's popular H-28 design spoke to a generation. Boats.com (boat review). Dominion Enterprises. http://www.boats.com/news-reviews/article/herreshoff-heritage (accessed January 1, 2009).

Hauser, J. R., and D. Clausing. 1988. The house of quality. *Harvard Business Review* May–June: 63–73.

Herreshoff, L. F. 1943. How to build H-28 twenty-eight foot auxiliary cruising ketch. *The Rudder: The Magazine for Yachtsmen* 59 (1): 50.

King, L. 2007. Sky's CRM plans were too vague to blame us, says EDS. ComputerWorldUK. http://www.computerworlduk.com/management/it-business/services-sourcing/news/index.cfm?newsid=5867 (accessed January 6, 2009).

Konrath, J. Is your value proposition strong enough? Marketing Profs. http://www.marketingprofs.com/print.asp?source=/2/konrath1.asp (accessed August 28, 2008).

Mizuno, S., and Y. Akao, eds. 1994. *Quality function deployment: The customer-driven approach to quality planning and deployment.* Rev. ed. Trans. G. Mazur. Tokyo: Asian Productivity Organization.

Saaty, T. L. 1980. *The analytic hierarchy process.* New York: McGraw–Hill.

References

Chapter 4

Explore the Design Space

In architecting a new program all the serious mistakes are made on the first day.

R. J. Spinrad

Plan to throw one away. You will anyway.

F. P. Brooks, Jr.

If you rush too quickly toward system [design], you will miss out on creative ways to satisfy user goals more effectively than you might by using the obvious first choice.

Martin Fowler, 1997

Introduction

Ulrich and Eppinger (1995) list a number of common pitfalls that design teams experience:

- Consideration of only one alternative: The team begins detailed design work on the first design idea that came to them. There is little or no discussion on whether this is the best approach or even if there are other approaches.
- Driven by the most assertive team member: A member who is confident that he or she has fully grasped the problem and has identified the solution sways the team at every decision point. This is usually a case of overconfidence.
- Involvement of only one or two members: One or two members of the team are fully engaged in the design process. They drive the decision making. If asked privately, the other members will typically complain that their own ideas are not respected.
- Lack of commitment by others: The active and engaged members of the team complain that the rest of the team is not participating: The others just ask to be told what to do. The active members feel the burden of the project resting on them.
- Failure to look for useful concepts from other applications: The product the team develops looks very much like past generations of the same product. Even though new technologies (new materials, new energy sources, new design principles) are being developed and applied elsewhere, the team seems unable to realize the relevance of these technologies.

The problem is that the team has not really looked outside the narrow domain of an existing product and its direct competitors.

■ Failure to consider entire categories of solutions: Again, the product looks like past generations of the same product. The team never seriously considers any radically different approaches. As the quote at the beginning of the chapter states, they miss the opportunity to satisfy user goals more creatively.

■ Ineffective integration of promising ideas: The team may consider a radically different idea, but it abandons it because it does not fit with how the team conceives of the rest of the design. It misses the fact that a radical idea in one area of the design will require new ideas throughout the design before it can be fully integrated. For example, one cannot move from a quill pen dipped in ink to a pen with an onboard ink supply without also redesigning the nib of the pen.

All of these pitfalls of team performance share a common feature: The team was not effective at engaging everyone on the team in the search for creative, integrated design concepts. The truth of the opening quote ("all of the mistakes are made on the first day") becomes apparent only late in the design cycle, if ever. It is only with reflection that you trace the failure to develop an innovative design to the assumptions and decisions made at the very beginning.

You will want a different outcome from your design team experiences. If that is the case, you will need to insist that the team spend a considerable amount of time up front in exploration. It is only by exploring the design space that you can be confident that your design solution is as creative as you first thought it to be. More than likely, the exploration process will uncover a better, more creative approach. Furthermore, through fanning out and actively looking for alternatives, you are likely to engage the most team members. The breadth of alternatives considered will not come from a single individual. The most fruitful idea may come from the least assertive team member. You create a climate for creativity by building times of exploration into your project plan and by recognizing when the design process should be interrupted with a new exploration phase.

Creativity is normally associated with an unstructured way of thinking. We speak of "creative leaps of imagination," the bridging of disparate thoughts. It would seem, therefore, that it is futile to talk about "steps to creativity," as though it were a mechanistic process. There is truth to that, but it also true that we can set the stage for creativity to happen; we can assemble the actors, arrange the scenery, and create the setting. That is what this chapter is about. The techniques of this chapter are focused on two parts of the process:

1. Discover concepts relevant to the design problem at hand.
2. Combine the concepts and generate integrated solutions.

A Missed Opportunity of Historic Magnitude

By 1842, Charles Babbage, an English inventor and mathematician, had developed a detailed design of a mechanical device to execute arithmetical computations automatically. Babbage's "analytical engine," as he called it, used cams, wheels, levers, and gears to perform arithmetic calculations. The design of the analytical engine included storing numbers in memory and then feeding the numbers back into subsequent calculations. The sequence of those calculations would be controlled by punched cards daisy-chained together. Babbage even designed a mechanical printer that could print the results of the calculations in tabular form. Mathematician Ada Lovelace,

Figure 4.1 **Experimental portion of analytical engine, 1871. (Used with permission from the Science & Society Picture Library.)**

an admirer of Babbage, published the first computer program, a set of instructions that could be implemented on the engine to compute the Bernoulli numbers. She also envisioned many other uses for the analytical engine, such as the generation of musical scores (Menabrea and Lovelace, 1842).

Only a portion of the engine was built before Babbage's death in 1871 (Figure 4.1). If he had been able to build his analytical engine, it would have been the world's first programmable computer. Unfortunately, Babbage's invention was viewed as an impractical novelty and his work slid into obscurity. The re-invention of the computer came over a century later with a new theory of computation and a new design approach based on electronics. The tremendous advance in computing in the latter half of the twentieth century created a resurgence of interest in Babbage's ideas. Eventually, in 2002, a team at the London Science Museum completed Babbage's difference engine #2 (essentially the analytical engine without punched cards). Their construction was faithful to Babbage's designs, and it performed flawlessly. The machine, on display in the London Science Museum, consists of 8,000 parts, weighs 5 tons, and measures 11 ft long. It is powered by a hand crank (Computer History Museum, 2008).

What prevented the construction of the analytical engine in Babbage's lifetime? Certainly part of the reason was the difficulty of machining and assembling a mechanical device of such complexity and precision using only the technology of the day. Politics, personality, and finances were also involved. Babbage had spent his fortune and considerable government funds on a precursor device, the "difference engine," but it was never completed. That project was cancelled by the British Parliament in 1839 and with it died the chance of funding development of the more advanced analytical engine.

It is intriguing to ponder the possibility that the most significant invention of the twentieth century, the computer, could have been developed in the middle of the nineteenth century. Clearly, the concept of a computer and a theory of computation were available. What was missing, perhaps, was the application of a more appropriate technology. Ralph Merkle, former professor at Georgia Tech College of Computing, believes that such a technology was available in Babbage's day: electromechanical relay switches. Relays were available in the 1830s and by the 1840s they were in widespread use in the nascent telecommunications industry.

Had Babbage and others explored the design space for "analytical engines," they might have realized that a relay-based computer was relatively easy to build and quite practical. But they did not, and so missed an opportunity of historic magnitude (Merkle 2003).

Discover Concepts

The first phase of the exploration process is to discover concepts that are relevant to the design problem. As with most discovery processes, it can be approached systematically. Following Ulrich and Eppinger (1995), we recommend the following steps:

1. Clarify the problem and decompose the functions.
2. Brainstorm and research.
3. Organize concept fragments.
4. Prune and expand.

Clarify the Problem and Decompose the Functions

We cannot speak about a creative solution without first perceiving the problem. This may seem like a trite statement, but we have observed teams expending tremendous energy on a project and yet achieving little because they lack a focus for their efforts. Everyone is doing something that contributes, but the objectives are so vague that there is no way to integrate the results. That is why requirements-driven design is so valuable. The requirements form a natural starting place for the design effort.

Table 4.1 is an extract of the functional requirements for the primary functionality of the toy catapult. These requirements define the central design problem. We must solve this problem in a way that satisfies all of the other requirements (safety, reliability, repairability, etc.), but the other requirements are not the starting point. That is, we can distinguish between requirements that specify primary capabilities and those that specify constraints. A *primary capability* is a function that the system must perform to accomplish the mission. A *constraint* is a requirement on how the system must perform its primary functions. In other words, we have contrasted the "what" (what the system must do) with the "how" (how it must do it).

We think it is best to approach primary capability requirements from a behavioral perspective. To do this, we return to the initial behavioral description (use case) that we developed for the toy catapult in Table 2.9. This time, however, we try to analyze the functionality into greater detail. Table 4.2 is a restatement of the behavior of the "child plays with toy" use case but with the functions decomposed into finer detail. Observe that we state these functions using abstract language so that we do not trap ourselves into a particular solution approach. We speak of "storing and accepting energy," "detecting a command to release," "converting stored energy to translational energy," and "applying translational energy." This is not just fancy language to obscure what is going on. It is a conscious attempt to state problems abstractly so that we have the greatest scope for thinking of solutions.

Table 4.1 Review Functional Requirements for Primary Capability

Index	*Originating Requirements*	*Abstract Function Name*
OR.1	The system shall detect the command to arm.	Detect
OR.2	The system shall secure its receptacle in the armed position. The force required to arm the system may be supplied by the system (e.g., battery power) or by the child.	Arm
OR.3	The system shall hold the projectile in its receptacle.	Load
OR.4	The system shall detect the command to release from a passing toy (lateral direction) and also from a child or falling projectile (vertical direction).	Trigger
OR.5	The system shall be capable of ejecting lightweight balls and dolls from its receptacle.	Eject

Having developed a detailed functional view of the problem, we next need to isolate those details that matter most. On any project, we will have limited time for exploration and innovation. It is important to focus the development effort on those aspects that have the greatest potential for innovation and meeting customer needs creatively. In Table 4.3, we highlight (bold italics) those functions that seem to us to have the potential of leading to fundamentally different design approaches. The functions that are not highlighted seem like they can be safely left to a later stage of the design process. For example, holding the projectile in the receptacle seems like a simpler design problem than the ability to detect the release command from two different directions.

The need to focus comes about because design spaces are so large. The team can easily become overwhelmed by all the design possibilities. We want to avoid the pitfalls mentioned at the beginning of the chapter that are caused by consideration of a design space that is too small. On the other hand, we recognize the constraints on the time and effort that are available.

Brainstorm and Research

By stating the problem in abstract functional terms and by focusing the development effort on the most difficult design problems, we have set the stage for the team to discover alternative ideas that address these problems.

A *concept fragment* is an idea that can be used to satisfy a functional requirement. It may or may not be practical, cost effective, or compatible with other ideas or other functional requirements. It is simply an idea that solves one aspect of the overall problem.

Brainstorming—the free expression of marginally feasible solutions—can be useful. For brainstorming to be effective, there must be agreement and discipline not to discuss the flaws in any proposed ideas until the brainstorming session is over. Without this discipline, social inhibitions will cut off the creativity. Keep the atmosphere light. Remember that, at this point, you are seeking creative solutions, rather than realistic solutions. It is sufficient if an idea solves only one aspect of a problem. Make lists of the ideas that surface during brainstorming. Stop the brainstorming session before the group tires of it. Move to a serious discussion of the ideas proposed.

Table 4.2   Decompose Functions Using Abstract Language

Child Plays with Toy		
Initial conditions		
1. System is in the unloaded state.		
Operator (child)	*System (toy)*	*Projectile*
The child commands the toy to arm		
	The system shall detect the command to arm	
	The system shall store or accept external energy	
	The system shall secure its receptacle in the armed state	
The child loads the receptacle		
	The system shall hold the projectile in its receptacle	
The child triggers the release		
	The system shall detect the command to release (laterally or vertically) from the child	
	The system shall convert energy to translational energy	
	The system shall apply translational energy to the receptacle	
	The system shall eject the contents of the receptacle	
		The projectile flies through the air and lands some distance away
Ending conditions		
1. The toy is in the unloaded state		

Table 4.3 Focus the Development Effort

Operator (child)	System (toy)	Projectile
The child commands the toy to arm		
	The system shall detect the command to arm	
	The system shall store or accept external energy	
	The system shall secure its receptacle in the armed state	
The child loads the receptacle		
	The system shall hold the projectile in its receptacle	
The child triggers the release		
	The system shall detect the command to release (laterally or vertically) from the child	
	The system shall convert energy to translational energy	
	The system shall apply translational energy to receptacle	
	The system shall eject the contents of the receptacle	
		The projectile flies through the air and lands some distance away

Note: Italics indicate functions that have the potential of leading to fundamentally different design approaches.

The goal in this phase of the discovery process is to uncover as many concept fragments as possible. This phase demands the maximum creative involvement from the team members. Brainstorming sessions can be fun as team members think of offbeat ideas that make the team laugh. Of course, if the leader finds that too many offbeat ideas are being generated, he or she may want to remind the group of other requirements that will need to be satisfied. In general, however, it is better to resist the urge to reject ideas until there are more than 10 or 20 ideas from which to choose.

In addition to brainstorming, concept fragments can be discovered through a process of research. Studying competitors' products will yield many relevant ideas. That is why it is important for designers to attend trade shows and competitive events. The search can go further afield than just competitive products. Are there *related products* that perform a similar function? For example, when considering toy catapults, it may be relevant to look at tennis ball machines and other ball-pitching machines to see how they are constructed. An Internet search for tennis ball machines reveals an interesting tennis ball launcher for exercising dogs (www.buygodoggo.com). *Patent searches* are another source of novel ideas. US Patent 5851012, for example, is a patent for a "ball game apparatus with spin imparting catapult." US Patent 4125106 is a patent for an "elastic powered catapult." *Trade publications and engineering handbooks* are another source of ideas.

Other ideas can come from talking with customers, particularly with customers who have special expertise or experience. In the case of toy catapults, toy aficionados, medieval warfare experts, and hobbyists can be consulted.

In Table 4.4, we used a brainstorming session to generate concept fragments for each of the three functions identified in Table 4.3. We could continue the process with a research phase, but this initial list is sufficient to illustrate the subsequent techniques.

Table 4.4   Generate Concept Fragments for Each Function

The System Shall Store or Accept External Energy	The System Shall Detect the Command to Release (Laterally or Vertically) from the Child	The System Shall Convert Energy to Translational Energy
Spring	Latch	Piston
Battery	Removable pin	Catapult (levered arm)
Flywheel	Motion sensor	Hammer
Human power	Proximity switch	Cannon
Wind	Gimbals	
Explosives	Vacuum (break seal)	
Compressed air	Wheel with protruding panes (horizontal and vertical)	
High-pressure fluid (hydraulics)	Pivot rings (interlocking rings)	
Internal combustion		
Solar electric cells		
Compressed carbon dioxide		
Electric motor		
Chemical reaction resulting in high-pressure gas		
Pendulum		

Table 4.5 Organize Concepts into Suggestive Categories (a Concept Classification Tree)

The System Shall Store or Accept External Energy						
Energy Source				Energy Storage		
Electrical	*Human*	*Chemical*	*Other*	*Mechanical*	*Pneumatic*	*Hydraulic*
Battery	Human power	Explosives	Wind	Spring	Compressed air	High-pressure fluid (hydraulics)
Solar electric cells		Chemical reaction resulting in high-pressure gas		Pendulum	Compressed carbon dioxide	
Electric motor		Internal combustion		Flywheel		

Organize Concept Fragments

A simple list of concept fragments is not as suggestive as a categorized list. The next step in the process is therefore to organize the list of concept fragments into a tree. A *concept classification tree* is an organization of concept fragments into a tree-structured hierarchy. The affinity process (Chapter 2) and the drag-and-drop technique in MS Excel can be used to group the concept fragments into categories and then group the categories. Table 4.5 and Table 4.6 show the categories that resulted from an affinity exercise applied in Table 4.4. Observe that we convey the same

Table 4.6 A Concept Classification Tree (Continued)

The System Shall Convert Energy to Translational Energy	The System Shall Detect the Command to Release (Laterally or Vertically) from the Child		
	Translate Direction		Secure and Release
	Mechanical	*Electrical*	
Piston	Gimbals	Motion Sensor	Latch
Catapult (levered arm)	Wheel with protruding panes (horizontal and vertical)	Proximity switch	Removable pin
Hammer	Vacuum (break seal)		
Cannon	Removable pin (in circular hole)		
	Pivot rings (interlocking rings)		

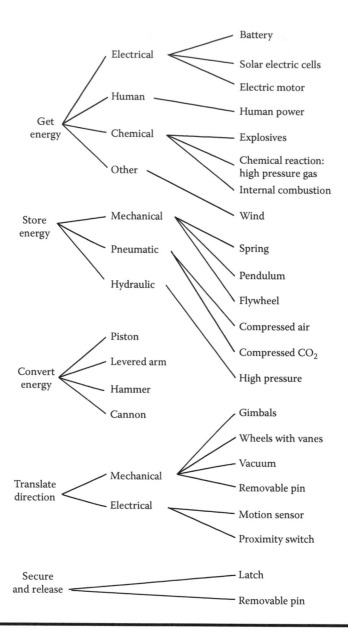

Figure 4.2 The concept classification tree in graphical form.

grouping information in Tables 4.5 and 4.6, using a tabular format with column headings, as we convey in Figure 4.2 using a graphical format.

Prune and Expand

The purpose of creating the concept classification tree is to see if the categories created suggest any further ideas. One can study the tree and see whether some branches should be expanded.

For example, we note that, under electrical detection methods, we list only motion sensors and proximity switches. We may choose to expand the branch of motion sensors and look for different types of sensors (photoelectric, infrared, radar, etc.). Another branch we might expand is the one for "human energy" so that we can see the diversity of human movements that could be exploited.

On the other hand, this is also an appropriate time to prune or eliminate branches from further consideration. For example, the notion of explosives surfaced during brainstorming. We kept the concept fragment in the mix until now to see if it would lead to something practical. Now, however, we have decided to prune the chemical branch of energy sources because of a concern for safety, maintenance, and operating cost. Table 4.7 and Table 4.8 show the pruning and expanding of the concept classification tree for the toy catapult.

Observe that we have decided to prune electrical sources of energy. This was done because the customer analysis revealed negative attitudes toward toys that require battery replacements ("avoid making a battery-powered toy"). However, there is a danger that we have pruned this branch too early. It could be that in spite of the customer attitude toward batteries, the most creative solution to the complete set of requirements involves an electrical power source. For example, by eliminating the electrical source of energy, we are also eliminating motion sensors. These sensors might be the ideal trigger mechanism. On the other hand, because the toy catapult example is simply for illustration; we need to constrain the design space.

Table 4.7 Prune or Expand Concept Fragments

The System Shall Store or Accept External Energy						
Energy Source				*Energy Storage*		
Electrical	*Human*	*Chemical*	*Other*	*Mechanical*	*Pneumatic*	*Hydraulic*
~~Battery~~ (Pruned cell)	Push	~~Explosives~~	~~Wind~~	Spring	Compressed air	~~High-pressure fluid (hydraulics)~~
~~Solar electric cells~~	Pull (Expanded cell)	~~Chemical reaction resulting in high-pressure gas~~		Pendulum	~~Compressed carbon dioxide~~	
~~Electric motor~~	Lift	~~Internal combustion~~		~~Flywheel~~		

Notes:

1. Electrical and chemical energy branches and electrical sensors branch pruned because of requirement for long time until first maintenance and low operating cost.
2. Other energy branch pruned because it was impractical.
3. Human power fragment expanded to distinguish between pushing and pulling actions.
4. Flywheel, hydraulics, hammer, gimbals, and vacuum fragments pruned because of high anticipated complexity or cost.

Table 4.8 Prune or Expand Concept Fragments (Continued)

The System Shall Convert Energy to Translational Energy.	The System Shall Detect the Command to Release (Laterally or Vertically) from the Child.		
	Translate Direction		Secure and Release
	Mechanical	Electrical	
Piston	~~Gimbals~~	~~Motion sensor~~	Latch
Catapult (levered arm)	Wheel with protruding panes (horizontal and vertical)	~~Proximity switch~~	Removable pin
~~Hammer~~	~~Vacuum (break seal)~~		
Cannon	Removable pin (in circular hole)		
	Pivot rings (interlocking rings)		

Notes:

1. Flywheel, hydraulics, hammer, gimbals, and vacuum fragments pruned because of high anticipated complexity or cost.
2. Electrical sensors pruned because of decision to prune electrical energy source.

Explore Concepts

The concept fragments are the raw material for generating design ideas. What is needed next is a systematic way to explore the multitude of ways in which these ideas can be combined to form design solutions. The following three steps are a natural progression for this process:

1. Combine the concept fragments.
2. Generate integrated concepts.
3. Identify the subsystems.

Combine Concept Fragments

A *concept combination table* (also called a *morphology box*) is matrix of concept fragments organized by function (each function forms a separate column) so that an integrated concept can be created by choosing any concept fragment for each function and combining them to satisfy the complete required functionality. Table 4.9 shows two concept combination tables: one for combining different ideas for the function "store or accept external energy" with ideas for "convert energy to translational energy" and another for combining ideas for "translating direction" with ideas for "secure and release" to perform the overall function "detecting the command to release." Below the concept combination tables, we have listed some of the different combinations that are suggested by these tables.

Note that we have chosen not to pursue all possible combinations. For example, the concept of a cannon works only with the compressed air concept. We also note that many, if not all, combinations of trigger mechanisms will work with the energy mechanisms. It is not useful to consider all possible combinations of trigger mechanism with energy mechanism. It will be better to defer integrating these concepts until we have narrowed our choices in each category.

Table 4.9 Create Concept Combination Tables

Concept Combination Table	
The System Shall Store or Accept External Energy	The System Shall Convert Energy to Translational Energy
Spring	Piston
Pendulum	Catapult (levered arm)
Compressed air	Cannon

Concept Combination Table	
The System Shall Detect the Command to Release (Laterally or Vertically) from the Child	
Translate direction	*Secure and release*
Wheel with protruding panes (horizontal and vertical)	Latch
Removable pin	Removable pin
Pivot rings	

Combine concepts	
The system shall store or accept external energy	The system shall convert energy to translational energy
Spring and piston	
Spring and catapult	
Pendulum and catapult	
Compressed air and piston	
Compressed air and cannon	

Combine concepts
The system shall detect the command to release (laterally or vertically) from the child
Wheel and latch
Wheel and removable pin
Removable pin
Pivot rings and latch

Notes:

1. All combinations of the table on the left with the table on the right are feasible.

It should be clear that every combination of ideas from a morphology box takes you down a different design path. It is a powerful mental tool for suggesting fresh approaches to design problems. Let us review the essence of the technique so far: Think about the problem from a functional perspective, analyze the overall function into distinct separate functions, think of different ways to solve each function, and systematically combine these alternatives to suggest new ways to solve the original problem.

Generate Integrated Concepts

The concept combination table is used to suggest different ways of approaching the overall design problem. The next step is to take each suggested approach (or, if time is limited, take the most

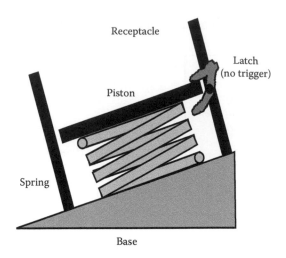

Figure 4.3   Spring and piston concept.

promising approaches) and work out an integrated design concept for that concept combination. This may simply be a sketch of how the ideas might work together or it may be a full-fledged design, perhaps to the level of building a prototype. It will be limited only by your skill and domain knowledge and by the time available.

This concept integration process must take place whether we are dealing with a mechanical design project, as with the toy catapult, an electrical design project, a software project, a clothing design project, or a service project. In the case of our toy catapult project, our mechanical engineering ability is limited. We have opted for simply sketching how the design concepts might be integrated. Figures 4.3–4.7 depict the five combination concepts for storing and converting energy.

For the spring and piston concept (Figure 4.3), a platform (square or circular) is attached to a rod (not shown) that can slide into a base to form a piston. A spring is trapped between the platform of the piston and the base. The piston and spring are surrounded by a box or tube that acts as a cannon. When the piston is depressed, there is room within the cannon to act as the receptacle for the projectile. Some form of a latch is used to catch the platform of the piston when it is fully depressed and hold it in the armed position. No trigger mechanism is shown.

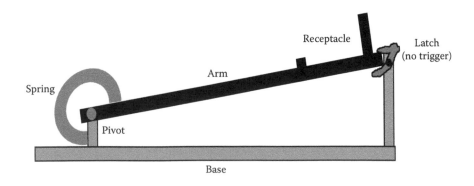

Figure 4.4   Spring and catapult concept.

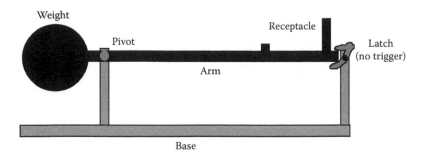

Figure 4.5   Weighted catapult concept (trebuchet).

For the spring and catapult concept (Figure 4.4), a long catapult arm with a receptacle attached is connected to the base at one end but allowed free movement around a pivot. A torsion spring (metal, rubber, or string based) is wrapped around the catapult arm at the pivot point and attached to the base. When the catapult arm is depressed, the tension in the spring increases. A latch catches the free end of the catapult arm and secures it in the armed position. No trigger mechanism is shown. When the latch is released the tension in the spring is released and the arm pivots upward. A crossbar of some sort (not shown) stops the arm at the desired angle and the projectile is launched.

For the weighted catapult concept (Figure 4.5), a catapult arm with an attached receptacle is connected to the base at a pivot point, but in this case, a portion of the arm extends beyond the pivot point and is attached to a pendulum counterweight. When the catapult arm is depressed, the counterweight is lifted, storing potential energy. As with the spring and catapult concept, a latch catches the free end of the catapult arm and secures it in the armed position. No trigger mechanism is shown. When the latch is released, the pendulum counterweight is free to fall. The arm pivots upward. A crossbar of some sort (not shown) stops the arm at the desired angle and the projectile is launched. This style of catapult is known as a trebuchet.

For the piston and compressed air concept (Figure 4.6), a circular platform is attached to a rod that can slide into a base to form a piston. The piston is enclosed in a tubular container to form a cannon. When the piston is depressed, there is room within the cannon to act as the receptacle for

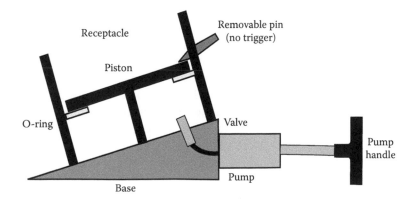

Figure 4.6   Piston and compressed air concept.

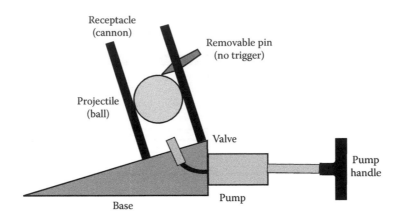

Figure 4.7   Cannon and compressed air concept.

the projectile. A rubber O-ring attached to the piston creates an airtight seal between the piston and the walls of the cannon. A removable pin is inserted through a hole in the wall of the cannon to catch the piston in the down position. Compressed air is introduced into the airtight cavity of the cannon below the piston through a valve at the base of the cannon. The concept shown in the figure has this compressed air coming from a hand pump. It is also possible to imagine the source being a purchased canister of compressed air or carbon dioxide. When the pin is removed, the pressure of the compressed air forces the piston to rise to its maximum position and launch the projectile. Residual compressed air escapes through the hole for the removable pin. No trigger mechanism is shown.

The cannon and compressed air concept (Figure 4.7) consists of a tube designed to snugly hold a standard ping-pong ball. When it is inserted into the cannon, the ping-pong ball creates an airtight cavity below it. The ball can be held in position by a removable pin and the rest of the concept follows the piston and compressed air concept.

Figures 4.8–4.11 depict four different combination concepts we developed for triggering the mechanism.

In Figure 4.8, the downward motion of the catapult arm catches the bottom of a latch (a). The latch pivots counterclockwise, pushing a catch rod out of the way. At a certain point in the rotation, the shape of the latch permits the catch rod to fall back. A spring attached to the catch rod ensures that the rod returns to its rest position near the anchor stop (b). At this point in the rotation, the top of the latch has caught the catapult arm and the catapult arm is secured. Its upward pressure is forcing the latch to turn in a clockwise direction (b), but the catch rod and trigger pin prevent the latch from turning. The pin is cone shaped and fits snugly into a hole in a side wall. Because of its conical shape, it can be knocked out of the hole easily with a force from any direction (horizontal or vertical). In the figure, it is shown as having a pull ring, but this could be replaced with a trigger arm. When the pin is removed, the catch rod is no longer pushed against the latch. It is free to drop out of the way and release the latch. When the latch is released, the upward force of the catapult arm causes it to rotate to the point where the catapult arm is released (c).

In the wheel and fin concept (Figure 4.9), two disks are connected to each other on a central axle (not shown). The axle penetrates the side wall of the catapult (not shown) so that one disk is inside the catapult and the other disk is outside. The axle is free to rotate within the hole in the side wall, so the pair of disks can spin freely. A weight attached to the inner disk causes the disks

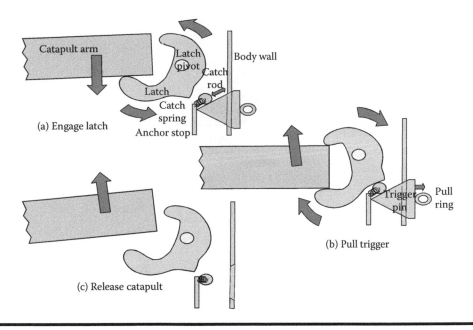

Figure 4.8   Latch and pin concept.

to be at rest in only one position. A catch shield is attached to the inner disk and projects into the interior of the catapult. When the pair of disks is at rest, the catch shield is at the top of its rotation cycle. The catapult arm has a pin that protrudes from its end. The catch shield on the inner disk catches this pin when it is in the at-rest position (a). The outer disk is equipped with fins that project outward. If either of these two fins is pushed, the pair of disks will rotate out of the at-rest position (b). In this case, the catch shield will also rotate and release the catch pin on the catapult arm, thereby releasing the catapult. The weight on the inner disk will then cause the pair of disks to return to the at-rest position. The fins can be oriented to be triggered by a falling object (the horizontal fin) or a passing toy train (the vertical fin).

In the clothespin and pull string concept (Figure 4.10), a compressive spring forces a catch lever to be at rest in the open position, like a clothespin. When the catapult arm is depressed, it pushes past the bottom of the catch lever, compressing the spring, Once the arm is past the bottom of the lever, the latch snaps open and the catapult arm is now secured (a). The upward pressure of the catapult arm is blocked by the catch lever. The catch lever can be pulled by a string, causing the compression spring to be compressed and the catapult arm to be released (b). The string passes through a hole in the side wall and is attached to a semispherical cup. When the compression spring is at rest, the catch lever is in the open position and the cup is pulled tightly against the wall (a). A trigger arm on the cup can cause the cup to rock in any direction (horizontal or vertical). When the cup is rocked, it pulls the string (b) and thereby releases the catapult arm.

Finally, Figure 4.11 displays the pivot ring concept. In this concept, two eye rings interlock and permit free movement, within limits, in any direction (pitch, yaw, and roll). One eye ring is attached to the base of the catapult and is fixed. The other eye ring is attached to a latch and is free to move. The latch is pulled upward by an expansion spring. As the catapult arm is pushed downward, the latch is pushed out of the way (the free eye ring rotates about the fixed eye ring).

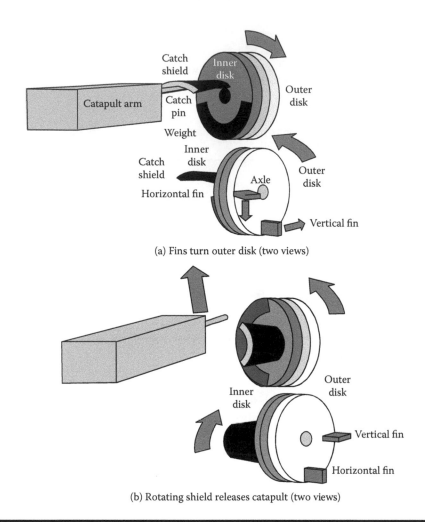

(a) Fins turn outer disk (two views)

(b) Rotating shield releases catapult (two views)

Figure 4.9 Wheel and fin concept.

When the arm passes the latch, the spring pulls it back into position. When the arm is released, it catches on the latch and is secured. A trigger arm protrudes from the latch and can be pushed in any direction (horizontal and vertical). The arm forces the moveable eye ring to rotate or twist around the fixed eye ring. This causes the latch to move out of the way and to release the catapult arm. The spring pulls the latch back into the resting position.

What we hope is apparent from this example is that there is a systematic process at least to set the stage for creative design. We have applied the process to a mechanical engineering problem, but it works just as well in software engineering, civil engineering, and other disciplines. Review the team dysfunctions that were listed at the beginning of the chapter and see that the process we have outlined and illustrated addresses all of the pitfalls. In particular, the technique encourages the search for numerous alternatives. It has a good chance of engaging the entire team. It explicitly looks for concept fragments from diverse sources. It also looks at the possibility of radical changes in multiple areas at once, thereby increasing the likelihood of discovering novel integrated solutions.

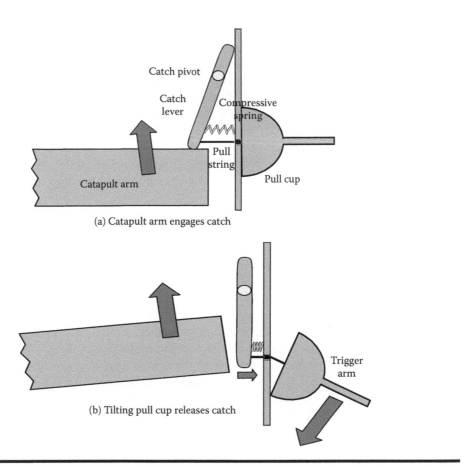

Figure 4.10   Clothespin and pull string concept.

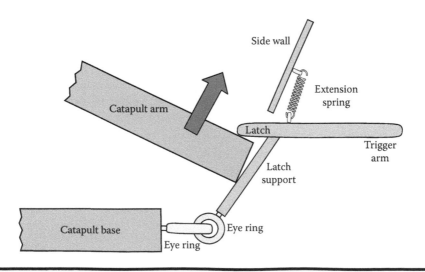

Figure 4.11   Pivot ring concept.

Identify Subsystems

By this stage in the process, we have created a set of integrated design concepts. Before we move into concept selection (the next chapter), it is useful to perform a summarization step. The appropriate summarization step at this stage is to identify and label the subsystems of our product. A *subsystem* is a collection of elements of a system that has an identifiable function of its own.

We have resisted identifying subsystems until this point out of a concern that we were not yet thinking about our system in broad enough terms. The concern is that if you identify subsystems too early in the design process, you become trapped into thinking about the design problem using your first design concept. Now that we have generated a number of very different integrated design concepts, that is no longer a concern.

We think of this step as part of a dive-and-surface cycle. We have just completed a dive phase in which we have generated very detailed integrated concepts for the central functionality of the product. Along the way, we have identified numerous components, such as springs, pins, and O-rings, that might be useful in our final design. We now enter the surface phase in which we look at all of the detail we have created and organize it into more abstract categories. For this phase, we use the affinity process and the drag-and-drop capabilities within MS Excel. This technique has been described earlier (in Chapter 2), so we simply show the input and output of this process for the toy catapult example.

For the input to the process, we create a long unordered list of the functions and components that we have been working with during the exploration of design concepts. Table 4.10 lists the functions with which we started and collects all the components that were named in our integrated concept diagrams. Each function and component occupies a separate cell in a spreadsheet.

The next step is to drag and drop these cells into new columns. We create a new column whenever a cell does not seem to fit ("have an affinity for") cells in an existing column. Note that we end by listing components from different integrated concepts together, provided they perform a similar function. It is a useful exercise to look at each component that has been identified and classify it by function. Finally, we look at the functionality expressed by the column and devise a column heading that captures that functionality. These column headings describe what we can now refer to as the subsystems of our product. Table 4.11 illustrates for the toy catapult. As a result of this grouping, we have identified five subsystems of the toy catapult: the energy storage system, the locking system, the projectile containment system, the trigger system, and the projectile launch system.

It is possible in this process of abstraction—that is, in listing subsystems—that you may discover a function that you had not thought about when you started the concept generation process. For example, after studying these integrated concepts, we see that the trigger function really comprises two functions: sensing and activating. When this happens, you may want to repeat the concept generation process by expanding the concept classification tree to explore this new functionality.

The dive-and-surface process is a creative activity that allows you to take a fresh look at system architectures. You can recombine the elements of design in creative new ways. Notice that we did not begin our design activity by listing subsystems. We waited until we had some different integrated concepts to work with. We then used these concepts as the raw material for thinking about subsystems.

General Motors Concept Car, "Hy-Wire"

If you begin your design process by listing subsystems, you will tend to fall into conventional patterns of design. The major components of a conventional automobile are shown in Figure 4.12. The example of GM's "AUTOnomy" and "Hy-Wire" concept cars (Figure 4.13 and Figure 4.14, respectively) shows what is possible when you take a fresh start.

Table 4.10   Collect Functions and Components

The system shall detect the command to arm
The system shall store or accept external energy
The system shall secure its receptacle in the armed state
The system shall hold the projectile in its receptacle
The system shall detect the command to release (laterally or vertically) from the child
The system shall convert energy to translational energy
The system shall apply translational energy to receptacle
The system shall eject the contents of the receptacle
Receptacle
Latch
Base
Arm
Pendulum weight
Valve
Pump
Pump handle
Removable pin
Piston
O-ring
Spring
Arm
Pivot
Trigger
Latch pivot
Catch rod
Catch spring
Trigger pin
Pull ring
Anchor stop
Catch lever
Catch pivot

(*Continued*)

Table 4.10 Collect Functions and Components (Continued)

Compressive spring
Pull string
Trigger arm
Pull cup
Inner disk
Outer disk
Horizontal fin
Vertical fin
Catch pin
Catch shield
Axle
Weight
Eye ring
Latch support
Extension spring

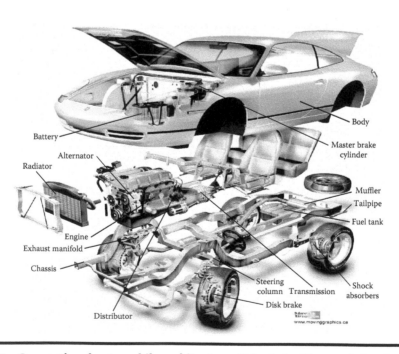

Figure 4.12   Conventional automobile architecture. (Adapted with permission from MovingGraphics.ca)

Table 4.11 Organize Functions and Components into Subsystems

Energy Storage System	Locking System	Projectile Containment System	Trigger System	Projectile Launch System
The system shall store or accept external energy	The system shall secure its receptacle in the armed state	The system shall hold the projectile in its receptacle	The system shall detect the command to release (laterally or vertically) from the child	The system shall convert energy to translational energy
Pendulum weight	Latch	Receptacle	Trigger	The system shall apply translational energy to receptacle
Valve	Removable pin		Trigger pin	Arm
Pump	Latch pivot		Pull ring	Piston
Pump handle	Catch rod		Pull string	Pivot
O-ring	Catch spring		Trigger arm	The system shall eject the contents of the receptacle
Spring	Anchor stop		Pull cup	
The system shall detect the command to arm.	Catch lever		Outer disk	
Base	Catch pivot		Horizontal fin	
	Compressive spring		Vertical fin	
	Inner disk		Axle	
	Catch pin		Eye ring	
	Catch shield			
	Weight			
	Latch support			
	Extension spring			

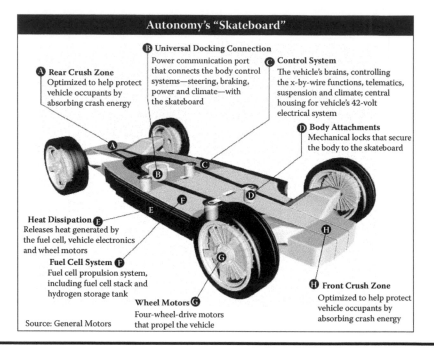

Figure 4.13 "Skateboard" architecture for General Motors concept car "AUTOnomy." (Used with permission from GM Media Archives.)

Figure 4.14 Sedan version of General Motors concept car "Hy-Wire." (Used with permission from GM Media Archives.)

Two basic elements largely dictate car design today: the internal combustion engine and mechanical and hydraulic linkages.

If you've ever looked under the hood of a car, you know an internal combustion engine requires a lot of additional equipment to function correctly. No matter what else they do with a car, designers always have to make room for this equipment.

The same goes for mechanical and hydraulic linkages. The basic idea of this system is that the driver maneuvers the various actuators in the car (the wheels, brakes, etc.) more or less directly, by manipulating driving controls connected to those actuators by shafts, gears, and hydraulics. In a rack-and-pinion steering system, for example, turning the steering wheel rotates a shaft connected to a pinion gear, which moves a rack gear connected to the car's front wheels. In addition to restricting how the car is built, the linkage concept also dictates how we drive: The steering wheel, pedal, and gear-shift system were all designed around the linkage idea.

The defining characteristic of the [General Motors] Hy-wire (and its conceptual predecessor, the AUTOnomy) is that it doesn't have either of these two things. Instead of an engine, it has a fuel cell stack, which powers an electric motor connected to the wheels. Instead of mechanical and hydraulic linkages, it has a drive-by-wire system—a computer actually operates the components that move the wheels, activate the brakes and so on, based on input from an electronic controller. This is the same control system employed in modern fighter jets as well as many commercial planes. ...

The result of these two substitutions is a very different type of car—and a very different driving experience. There is no steering wheel, there are no pedals and there is no engine compartment. In fact, every piece of equipment that actually moves the car along the road is housed in an 11-inch-thick (28 cm) aluminum chassis—also known as the skateboard—at the base of the car. Everything above the chassis is dedicated solely to driver control and passenger comfort.

This means the driver and passengers don't have to sit behind a mass of machinery. Instead, the Hy-wire has a huge front windshield, which gives everybody a clear view of the road. The floor of the fiberglass-and-steel passenger compartment can be totally flat, and it's easy to give every seat lots of leg room. Concentrating the bulk of the vehicle in the bottom section of the car also improves safety because it makes the car much less likely to tip over.

But the coolest thing about this design is that it lets you remove the entire passenger compartment and replace it with a different one. If you want to switch from a van to a sports car, you don't need an entirely new car; you just need a new body (which is a lot cheaper).

Harris (2002; reprinted with permission from HowStuffWorks.com)

Summary

In this chapter, we paused to explore other possible solutions to our design problem. We noted that many common pitfalls that design teams fall into can be traced to an unwillingness to explore the design space. We began by reviewing our functional requirements and selecting a small number of functions that would have the greatest impact on our design. We attempted to state these functional requirements abstractly so they would admit many possible solutions. We used the technique of brainstorming to generate a large number of concept fragments: potentially useful but isolated ideas. We noted that research into related products, patents, and trade journals, as well as talking with expert users, could also lead to design ideas. We used the affinity process to organize these concepts into a concept classification tree.

We then studied the tree and we pruned branches that we did not believe would lead to designs that would satisfy all our requirements. We also expanded branches that had not received enough attention during the brainstorming session. We then created a concept combination table to explore possible combinations to satisfy our different functional requirements. Selecting one concept fragment from each functional requirement column resulted in a phrase that would suggest an integrated concept. From these phrases, we generated sketches of integrated concepts.

From the integrated concept sketches, we collected an unordered list of functions and components. Using the affinity process, we classified the elements of this list into basic abstract subsystems of our design. We used the General Motors concept car Hy-Wire to illustrate that radically new architectures are possible when you take a fresh look at the design space.

The dive and surface metaphor was used twice in this process of exploring the design space: organizing concept fragments and summarizing the subsystems.

Discussions

1. Identify some likely consequences had a relay-based programmable computer become available in the middle of the nineteenth century.
2. How has GM's skateboard concept, first introduced with the AUTOnomy, influenced subsequent alternative-fuel vehicle designs?

Exercises

1. The goal of this exercise is to have you go through all of the steps of the concept generation process. This is a creative exercise, so you will have to contribute most of the thought. You may choose any design problem to be the basis for your concept generation. It may be best to choose a problem within your domain expertise (engineering, software, service industry, household utensils, etc.). If you want to continue working with the bathroom cleaning robot, then I would suggest that the problem of dealing with the toilet seat cover and toilet seat pose some special challenges, even if you have a flexible robotic arm at your disposal. How many different ways can you think of to solve the problem? (The simplest might be to sound a buzzer and wait for the cleaning attendant to lift the cover or the seat for you.) The problem you choose should have at least two related functional requirements so that you can create a morphology box and consider different integrated solutions. You should download and use the spreadsheet "Toy04 Concept Generation.xls" as a model for the different steps. Your solution should show evidence of the following steps:
   a. function decomposition;
   b. focusing development efforts;
   c. brainstorming (or researching) to generate concept fragments;
   d. organizing concepts into suggestive categories;
   e. pruning or expanding concepts;
   f. creating a concept combination table (a morphology box);

g. combining concepts—that is, listing the combinations that might work;

h. generating at least two integrated concepts (you may describe your concepts in words or sketch them); some real components should be part of this design; and

i. using the affinity process to classify the components from all the integrated concepts by function; naming the resulting subsystems in a generic fashion.

You do not have to select which concept is best at this stage. That is the goal of the concept selection process, which comes next.

References

Brooks, F. P., Jr. As quoted in Maier and Rechtin, 1997.

Computer History Museum. 2008. The Babbage engine. http://www.computerhistory.org/babbage/ (accessed December 11, 2008).

Fowler, M., with K. Scott. 1997. *UML distilled: Applying the standard object modeling language.* Addison–Wesley Object Technology Series. Reading, MA: Addison–Wesley.

Harris, T. 2002. How GM's Hy-Wire works. 17 October. HowStuffWorks.com. http://auto.howstuffworks.com/hy-wire.htm (accessed December 15 2008).

Maier, M. W., and E. Rechtin. 1997. *The art of systems architecting.* New York: CRC Press.

Menabrea, L. F., and A. Augusta (Lovelace). 1842. Sketch of the analytical engine. Bibliothèque Universelle de Genève, October, No. 82. http://www.fourmilab.ch/babbage/sketch.html (accessed December 11, 2008).

Merkle, R. 2003. Inside foresight: Focusing on assemblers. *Foresight Update* 51:2. (http://www.foresight.org/Updates/Update51/Update51.2.html (accessed December 11, 2008).

Spinrad, R. J. As quoted in Maier and Rechtin, 1997.

Ulrich, K. T., and S. D. Eppinger. 1995. *Product design and development.* New York: McGraw–Hill.

Chapter 5

Optimize Design Choices

My way [of making difficult decisions] is to divide half a sheet of paper by a line into two columns; writing over the one Pro, and over the other Con. Then during three or four days' consideration, I put down under the different heads short hints of the different motives, that at different times occur to me, for or against the measure. When I have thus got them all together in one view, I endeavor to estimate their respective weights: and where I find two, one on each side, that seem equal, I strike them both out. If I find a reason pro equal to some two reasons con, I strike out the three … and thus proceeding I find at length where the balance lies; and if, after a day or two of further consideration, nothing new that is of importance occurs on either side, I come to a determination accordingly. … I have found great advantage from this kind of equation, in what may be called moral or prudential algebra.

Benjamin Franklin, 1772

Introduction

The preceding quote from Benjamin Franklin is one of the earliest known mentions of a technique for trade-off analysis. It remains one of the most eloquent statements of how to approach a complex decision-making problem. There are numerous points in the design process when the designer must make a decision in order to proceed. We are at such a juncture now with the toy catapult example. Having generated a collection of integrated design concepts, we must select one of the design concepts to be the basis for our subsequent design efforts.

Many more decisions will need to be made, both artistic and engineering-related. As Franklin observed, these decisions are difficult because of the number of factors involved. One of the chief pitfalls in team-based decision making is the tendency to get locked into *circular discussions*. Someone will argue that design B is better than design A, someone else says that design C is better than design B, and a third person argues that design A is better than design C. There seems to be no way to break out of the loop. The problem, of course, is that everyone is comparing the designs according to different product objectives. Design B has the best performance, design C is the lowest cost, design A is the most reliable, and so forth. It can be a complex problem to consider all the designs and all the factors simultaneously and to give the proper weight, in the sense of Franklin's "prudential algebra," to each factor.

A *trade study*, sometimes called a trade-off study, is a formal consideration of the multiple factors involved in a selection choice and the rationale for selecting the best choice. There are a number of techniques for concept selection (such as Benjamin Franklin's pro and con list), but the most commonly cited method is the Pugh concept selection matrix, known as a *Pugh analysis*, after the Scottish professor, Stuart Pugh, who developed the methodology in the 1980s.

We encourage you to document your trade studies. A reasonable set of documents would include the problem statement, tables of raw data used as input for benchmarking the alternative designs, copies of the Pugh matrices used in the trade-off analysis, and a statement of the final decision. The documentation permits you to revisit a decision and quickly see if new information might change the best choice. It can also benefit others in your organization who are faced with similar design choices. Imagine how you might use a searchable database of trade studies.

Lifeboats on the *Titanic*

The original design for the *RMS Titanic* [Figure 5.1] called for 64 lifeboats, but this number was reduced to 20 before its maiden voyage. This was a trade-off mistake. The chief designer wanted 64 lifeboats, but the program manager reduced it to 20 after his advisors told him that only 16 were required by law. The chief designer resigned over this decision. The British Board of Trade regulations of 1894 specified lifeboat capacity: For ships over 10,000 tons, this lifeboat capacity was specified by volume (5,500 cubic feet), which could be converted into passenger seats (about 1,000) or the number of lifeboats (about 16). So, even though the *Titanic* displaced 46,000 tons and was certified to carry 3,500 passengers, its 20 lifeboats complied with the regulations of the time.

**Smith et al. (2007; reprinted with permission of Wiley–Blackwell, Inc.,
a subsidiary of John Wiley & Sons, Inc.)**

Figure 5.1 *RMS Titanic* **leaving Queenstown, Ireland, on April 11, 1912. (Courtesy of the Father Browne Collection and the** *Titanic* **Historical Society and** *Titanic* **Museum.)**

Select Concepts

Using the techniques presented in the previous chapter, it is likely that your team has generated numerous design concepts. We now use Pugh analysis (Pugh 1991) to select a single concept. Following the approach by Ulrich and Eppinger (1995), the steps in our version of Pugh analysis are to

1. Identify the alternative design concepts;
2. Identify the relevant attributes (product objectives);
3. Perform an initial screening of alternatives;
4. Rate the alternatives in each attribute;
5. Weight the attributes;
6. Score and rank the alternatives; and
7. Select an alternative.

We illustrate the technique by continuing with the toy catapult example.

Identify Alternatives

Table 5.1 lists the five integrated design concepts that we developed in the previous chapter for the launch subsystem. Our goal in this exercise is to select a single concept as the most promising design concept.

Identify Attributes

Table 5.2 contains all the customer product objectives that were identified during our customer analysis study. These are the dimensions or attributes along which we want to measure the integrated design concepts. If the design concepts differ in ways that do not affect any of these product objectives, then those differences are not relevant to the decision. Viewed another way, if we discover an important way in which the concepts differ, and it is not one of our product objectives, we may have overlooked an objective. In that event, we should revisit the set of product goals and test whether they are complete.

In the context of decision making, we refer to these product objectives as product decision *attributes*. Not all the product attributes need to be considered for every design decision. Considering the different design concepts and the parent's skill level, we conclude that the fun value for the parent is likely to be very similar for all the designs. We eliminate the "fun for parent" attribute from the list. We also eliminate the "appealing to child" and "appealing to parent" attributes because they are not relevant at this stage of design. Decoration, material choice, and personality are features that can be added later, irrespective of the particular design concept. Finally, we eliminate the repairability attribute for a similar reason. The designs probably do not differ much in their intrinsic repairability.

Table 5.1 Toy Catapult Design Concept Alternatives

Spring and piston	Weighted catapult	Spring and catapult	Compressed-air piston	Compressed-air cannon

Table 5.2   Toy Catapult Design Attributes (Customer Product Objectives)

Goals	Attribute Name
Make the toy fun for the parent	Fun for parent
Make the toy fun for the child	Fun for child
Make the toy appealing to the parent	Appealing to parent
Make the toy appealing to the child	Appealing to child
Make the toy safe	Safety
Make the toy easy to put away	Storability
Make the toy playable for a long time	Reliability
Make the toy easy to repair	Repairability
Make the toy affordable	Affordability

Even if they did, the property of repairability may be strongly correlated with affordability. Table 5.3 is the resulting list of attributes after we have eliminated attributes we deemed to be unnecessary.

Observe the principles under which you can safely *eliminate an attribute* from consideration: Either (1) the design alternatives do not differ from each other with respect to this attribute, or (2) the design alternatives differ in this attribute in a way that is strongly correlated with another attribute.

Screen the Alternatives

The next step is to perform an initial *screening of the concepts*. There are likely concepts in the list that are clearly not worth pursuing. It is not worth the effort of evaluating these concepts with the same level of attention that we will give to the serious candidates. The screening step uses a very simple rating and scoring system to identify those concepts that should be eliminated early in the process. Table 5.4 shows a screening matrix for the toy catapult design alternatives. The attributes of the decision are listed as rows. The design alternatives are columns. One design alternative is listed as the *reference concept*, which is the design alternative against which all other concepts are compared. It is easier to say that a design alternative is better or worse than another concept than

Table 5.3   Eliminate Unneeded Attributes

Goals	Attribute Name
Make the toy fun for the child	Fun for child
Make the toy safe	Safety
Make the toy easy to put away	Storability
Make the toy playable for a long time	Reliability
Make the toy affordable	Affordability

Table 5.4   Screen the Concepts Relative to a Reference Concept

Attribute Name	Spring and Piston (Reference Concept)	Weighted Catapult	Spring and Catapult	Compressed-Air Piston	Compressed-Air Cannon
Fun for child	0	0	0	–	–
Safety	0	0	–	0	0
Storability	0	–	–	–	–
Reliability	0	+	+	–	–
Affordability	0	+	+	–	–

it is to score each alternative on some absolute scale. It is best if the reference concept is chosen to be one of the designs likely to be a serious candidate for the final choice. We chose the "spring and piston" design to be the reference concept because it was actually the first design we thought of. The other designs resulted from our research and exploration phase.

For each cell in the resulting matrix, rate the alternative for that column against the reference concept for the particular attribute in that row. For screening, use a simple rating system: Use a zero if the concept is no better or worse than the reference concept, use a plus (+) if the concept is better than the reference concept in that attribute, and use a minus (–) if the concept is worse than the reference concept. The reference concept receives a zero in each row. Table 5.4 shows the results of our rating of these alternative concepts.

It is difficult to predict the fun factors of the different toy designs. We felt that the pumping action of the compressed air concepts would be difficult for a young child and would interfere with the fun. We also felt that the catapult concepts (levered arms) posed a greater safety risk because of the potential for the arm striking the face of the child. However, the weighted catapult has a slower action than the spring and catapult so it should be at least as safe as the reference concept. The reference concept has the most compact design possibilities. All of the other designs were rated lower in the storability category. The compressed air concepts depend critically on the achievement of a good seal: either through an O-ring or through a perfect fit of the projectile to the cannon. Wear or erosion of the O-ring or the projectile could cause these concepts to fail. We also rated the spring and piston (the reference concept) lower than the catapults because the forces on the piston in the armed position will likely cause faster wear on this part. In the affordability attribute, we estimated that the catapult designs would be the easiest to manufacture in a hobbyist's shop.

The advantage of the *screening matrix* is that it shows quite obviously which concepts are dominated. Observe that the compressed air concepts are not better than any of the other concepts in any attribute, and they are worse than the reference concept in several attributes. These are *dominated concepts*. Now is the logical time to cease development effort on these design concepts. Table 5.5 shows the concept screening matrix with the dominated concepts eliminated.

Someone on the design team may object that the flaws in a dominated design can be easily corrected. If that is the case, then a decision must be made as to whether it is worth the development time and effort to correct these flaws and, subsequently, to revisit the screening decision, or whether it is better to push ahead with the remaining concepts. These are difficult decisions.

Table 5.5 Eliminate Dominated Concepts

Attribute Name	Spring and Piston (Reference Concept)	Weighted Catapult	Spring and Catapult
Fun for child	0	0	0
Safety	0	0	–
Storability	0	–	–
Reliability	0	+	+
Affordability	0	+	+

Prof. Al George says a student on the FSAE race car team (Appendix A, "Case Study: Formula SAE Racing Competition") always pleads for "just another week" to get some unconventional idea to work. The danger is that week after week will go by and the whole design, build, and test schedule of the race car will be put into jeopardy while waiting for the student to be successful. Sometimes it is better to go with a conventional solution and stop the exploration of unconventional ideas.

Of course, if the reference concept is dominated, then it should be eliminated and a new reference concept chosen to replace it. To avoid this occurrence, we recommend choosing the reference concept in the first place from one of the more serious candidates.

It can also be useful at this stage to get a preliminary ranking of the remaining concepts. Compute a score for each design concept by adding up the pluses and minuses in each column. The reference column will always have a score of zero. Table 5.6 shows the scores for remaining concepts in the toy catapult example. The weighted catapult concept has a net score higher than the spring and piston concept. This was something of a surprise because, as mentioned earlier, the spring and piston concept was the first concept we thought of. It is now becoming evident that the exploration phase may be leading us to a better design.

Table 5.6 Rank the Remaining Concepts

Attribute Name	Spring and Piston (Reference Concept)	Weighted Catapult	Spring and Catapult
Fun for child	0	0	0
Safety	0	0	–
Storability	0	–	–
Reliability	0	+	+
Affordability	0	+	+
Net score	0	1	0
Rank	2	1	2

Table 5.7 Rating Scheme Relative to Reference Concept

Relative Performance	Rating
Much worse than reference concept	1
Worse than reference concept	2
Same as reference concept	3
Better than reference concept	4
Much better than reference concept	5

Rate Alternatives

The screening matrix has served the purpose of eliminating dominated designs, but it is too coarse a measurement scheme to assist in deciding between alternatives that are close in character. Furthermore, the scoring scheme in the screening matrix is too simplistic for making a final choice. We repeat the evaluation process for the remaining concepts using a more detailed rating scale and a more detailed attribute weighting scheme. To begin, we replace the simple plus–zero–minus rating system with a more finely graded scale. Table 5.7 shows a *five-point scale* commonly used for comparing alternatives to a reference concept. One could also go to a seven-point or a nine-point scale if the differences between concepts are subtler.

In Table 5.8 we have re-evaluated all of the remaining concepts against the reference concept using this more finely graded rating scale. Note that we have eliminated the reliability attribute. It was felt that, having eliminated some concepts for which reliability was a concern, reliability was no longer a distinguishing factor. For example, the piston mechanism could be reinforced with a tongue and groove to eliminate the reliability concern. We also generalized the spring and catapult concept to include a torsion-powered catapult such as the one considered in the benchmarking study (Chapter 3). With this assumption, the manufacturability of either of the catapult designs was felt to be much better than the spring and piston concept. Research into counterweight-powered catapults (the weighted catapult concept) revealed that, in general, they are capable of throwing much heavier projectiles than other catapults, but that they have shorter ranges and slower release velocities (Smith 2001). Because the fun of a catapult is likely to come from its range and release velocity, we scored the spring and catapult higher than the weighted catapult in the fun attribute. On the other hand, release velocity is one of our primary concerns for the safety of the child, so this explains the lower rating for safety of the spring and catapult concept.

Table 5.8 Toy Catapult Design Concept Ratings

Attribute Name	Spring and Piston (Reference Concept)	Weighted Catapult	Spring and Catapult
Fun for child	3	3	4
Safety	3	3	2
Storability	3	2	2
Affordability	3	5	5

Table 5.9 Toy Catapult Attribute Weights

Attribute Name	Weight
Fun for child	17%
Safety	20%
Storability	7%
Affordability	6%

Weight the Attributes

We have already conducted an extensive exercise in Chapter 3 to determine relative weights for the different product objectives. We are dealing with a subset of the product objectives here, so we simply extract the relevant weights from Table 3.15. Table 5.9 shows the resulting weights for the attributes of the concept selection problem.

Observe that the weights do not sum to 100% because we are looking at only a subset of the full set of product objectives. It would be easy to normalize these weights so that they sum to 100%, but this would add no value to the analysis.

These weights now enter the decision-making process in a critical way, so it is important to review the weighting process. At the time we developed the weights, we were considering the product objectives in the abstract. Now, with specific design alternatives before us, the objectives are not so abstract. For example, with a little more effort, we could estimate the different costs of these designs. Similarly, knowing these designs, we could estimate the minimum age of the child for whom each design would be considered safe. We could then create customer survey forms tailored to these ranges of cost, age level, and other attributes. That is, now that we know the likely ranges for certain product characteristics, we could do a better job of estimating relative weights of the corresponding product objectives. However, for the purpose of this text, we will continue using the weights listed in Table 5.9.

Score and Select Alternatives

We now bring together the attribute weights with the alternative ratings to form the final *Pugh concept selection matrix*. Table 5.10 shows the weights and the ratings, computes a weighted score for each concept–attribute combination, and a total score for each concept. As can be expected, the scores are quite close for the remaining concepts.

Numerically, the weighted catapult concept is marginally better than the other concepts. At this point, the design team must review the data to see if the ranking is in accord with its understanding of the design issues. Note that the discussion can focus on the matrix and avoid the trap of circular discussions. If someone on the team wants to argue that the spring and catapult is the better design, he or she must:

- argue that its rating in some category should be increased;
- argue that the weights on some attributes should be changed; or
- identify a dimension of comparison (a new attribute) that has not been considered.

Table 5.10 Toy Catapult Design Concept Scoring and Ranking

Attribute Name	Weights	Spring and Piston (Reference Concept)		Weighted Catapult		Spring and Catapult	
		Rating	Weighted Score	Rating	Weighted Score	Rating	Weighted Score
Fun for child	17%	3	0.51	3	0.51	4	0.68
Safety	20%	3	0.6	3	0.6	2	0.4
Storability	7%	3	0.21	2	0.14	2	0.14
Affordability	6%	3	0.18	5	0.3	5	0.3
Total score			1.5		1.55		1.52
Rank			2		1		3
Continue?			No		Develop		No

The matrix approach is a highly effective way to organize the discussion and bring the design team to a consensus. It is also a concise tool to document the trade study and to present the decision process to an external audience.

In the end, we decided to develop the weighted catapult concept. The true weights used in our decision placed an even greater emphasis on safety than did the weights used in the concept selection matrix. The slower release velocity of the weighted catapult versus the other designs seemed to offer an inherent safety advantage. We also valued manufacturability (affordability) considerably more than storability.

Observe how closely we have come to Benjamin Franklin's approach (the quote at the beginning of this chapter) to making a decision in complex circumstances. We have found a way to present all of the considerations on a single page. We have weighed the pros and the cons and cancelled aspects that seemed in balance. We have reflected upon the choices in their most critical attributes and have arrived at a decision that can be justified with easily communicated statements.

The *Titanic* Lifeboat Question Revisited

Prof. Eric Smith and his colleagues have created the following hypothetical trade-off study that could have been performed (but was not) in the decision to limit the number of lifeboats on the *Titanic* from 64 to 20:

> But let us go back to the design decision to reduce the number of lifeboats from 64 to 20. What if they had performed the following hypothetical trade-off study? [In Table 5.11], the weights of importance range from 0 to 10, with 10 being the most important and the evaluation data (scores) also range from 0 to 10, with 10 being the best. For simplicity, we have not used scoring functions, so the evaluation data are also the scores. The magnitudes of the program manager's and the chief designer's alternative ratings are not in themselves important. What is important is that they indicate different preferred alternatives, which result from different sets of weights of importance.

Smith et al. (2007; reprinted with permission of Wiley–Blackwell, Inc., a subsidiary of John Wiley & Sons, Inc.)

Table 5.11 An Apocryphal Trade-Off Study for the *RMS Titanic*

Criteria	Weights of Importance		Alternatives and Their Input Values			
	Program Manager's Weights	Chief Designer's Weights	10 Life-Boats	20 Life-Boats	30 Life-Boats	64 Life-Boats
Will it satisfy the Board of Trade regulations? (yes, no)	10	10	0	10	10	10
Amount of deck space required for the lifeboats (ft²)	2	2	10	8	4	2
Possible perception that the ship is unsafe caused by the presence of a large number of lifeboats	6	2	10	8	4	2
Cost to purchase, install, maintain, launch, and operate the lifeboats (£)	9	4	10	8	4	2
Percent of passengers and crew that could be accommodated in lifeboats, if all lifeboats were launched full of people	1	10	2	4	6	10
Final scores produced by summing the program manager's weights × scores			172	**240**	174	144
Final scores produced by summing the chief designer's weights × scores			100	204	192	**216**

Source: Smith, E. D. et al., 2007. *Systems Engineering* 10 (3): 236–237. Reprinted with permission of Wiley-Blackwell, Inc., a subsidiary of John Wiley & Sons, Inc.

Note: Best alternatives, based on weights, are shown in bold.

The point of this example is that conducting a trade-off analysis is no guarantee against bad decisions. Some flaws in decision making go much deeper than simply a failure to weigh all the factors. Discussion questions at the end of this chapter explore some of these flaws.

Optimize Parameters

In the previous sections, we have considered discrete design choices: a limited number of alternatives. But often the design decisions differ in degree rather than in kind. In this section, we consider optimizing *continuous design parameters* subject to engineering and design constraints.

Optimizing design parameters typically comes later in the design cycle than we are presenting here. It arises at the subsystem level when engineers are faced with manipulating subsystem-level

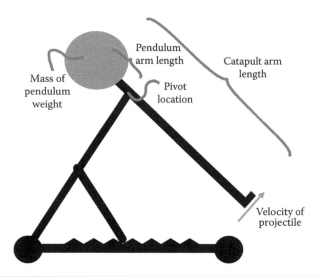

Figure 5.2 Elements of parameter optimization for catapult design.

engineering characteristics to meet system-level requirements. The example in this section focuses on the design of the catapult launch mechanism.

Based on the requirements analysis and the house of quality, we have requirements for the desired launch release velocity and launch mass (OR.14 and OR.15 in Table 3.19). At some point in the engineering design process, we will need to consider how the subsystems are going to satisfy these requirements. Based on the pendulum design, it is easy to see that several subsystem engineering characteristics will be involved. Figure 5.2 illustrates three of them: the mass of the counterweight, the length of the catapult arm, and the position of the pivot point. A more detailed analysis (see the appendix to this chapter) identifies three more: the mass of the projectile, the height of the counterweight above the pivot point prior to launch, and the height of the counterweight below the pivot point at the time of launch.

We assemble these parameters into a table (Figure 5.3), placing the system-level parameters in the rows and the subsystem-level parameters in the columns. This is the beginning of a house of quality at the subsystem level. It is called a *linked house of quality* because the rows of this table were the columns in an earlier house of quality (Figure 3.10, the house of quality for the toy catapult). Note that we have supplemented the table with the target technical performance measures from the earlier house of quality (requirements OR.14 and OR.15). These targets appear in a column on the right of the table. The imputed importance of each of these two technical performance measures (maximum launch velocity and maximum mass launch capability) was set at 90% (refer to the imputed importance row of Figure 3.16), so they both have equal relative weights (0.50) in the linked house of quality. The new subsystem engineering characteristics are in the columns. We have identified direction of change, units of measure, and relationships to the system-level performance measures for each of these subsystem characteristics.

The subsystem engineering characteristics can be viewed as *design parameters:* specifications that we have some freedom to change provided we satisfy system-level performance requirements. Observe that we have six design parameters and only two requirements to satisfy (OR.14 and OR.15). The difference between the number of design parameters and the number of requirements they must satisfy is called the *degrees of freedom*. In this example, we have four degrees of freedom.

	Relative Importance	Counterweight mass	Projectile mass	Length of catapult arm	Location of pivot from counterweight center of gravity	Beginning height of counterweight above pivot	Ending height of counterweight below pivot	Units	Targets
Direction of Change		↓	↑	↓	↑	↓	↑	Units	Targets
Maximum launch velocity	0.5	XX	XX	XX	?	XX	XX	mps	6
Maximum mass launch capability	0.5	XX	√√	XX	?	XX	XX	g	150
Measurement Units		g	g	m	%	m	m		

(System-Level Engineering Characteristics; Subsystem-Level Engineering Characteristics)

Figure 5.3 Linked partial house of quality for toy catapult.

When a design problem has degrees of freedom, it is usually possible to set the design parameters to satisfy the requirements, and to minimize or maximize some other objective, such as cost.

In addition, there are usually bounds or limits on how much some design parameters can vary. For example, the pivot point must be located somewhere on the catapult arm. Its location is constrained by the length of the catapult arm. Note that we have identified all the interactions between the system-level performance measures and the subsystem design parameters, except in the case of the pivot location (the question marks in the table). For example, reading the table, we say that "a decrease in the length of the catapult arm will strongly and negatively affect both the maximum launch velocity and the maximum mass launch capability." This should be obvious from a basic understanding of physics. Without experimentation or analysis, however, it is not easy to identify the impact of moving the location of the pivot.

We are led to formulate *optimization problems:* choose design parameters, within limits, to minimize (or maximize) some product objective subject to satisfying the design requirements. For example, the most expensive material in the toy catapult is likely to be the metal used to make the pendulum weight. Therefore, a simple example of parameter optimization is to choose the location of the pivot, for a given length of catapult arm, to minimize the mass of the pendulum weight. The constraints are that the combination of pivot location and pendulum weight must be sufficient to guarantee the required launch velocity and launch mass of the projectile.

At this point, you could construct prototype catapults and experiment with them to find the best design. This can be expensive in both time and materials because of the range and number of possible combinations of parameters. Formal techniques, called *design of experiments,* can be used to search this space of possibilities. Another approach is to use engineering analysis. The pivot

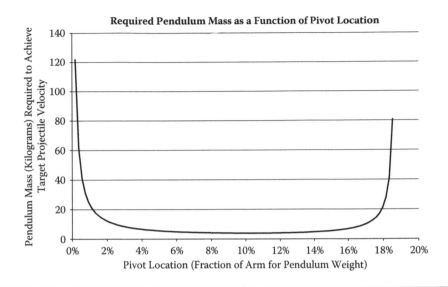

Figure 5.4 **Required pendulum mass as a function of pivot location.**

location problem can be posed as an elementary problem in physics and solved using a spreadsheet program. However, because this text is intended for nonengineers as well as engineers, we hide the details of the physics in an appendix to this chapter. The results of the analysis are two formulas: one for the required mass of the pendulum weight as a function of the pivot location (displayed in Figure 5.4) and another showing the projectile launch velocity as a function of the pivot location for a given pendulum weight (Figure 5.5).

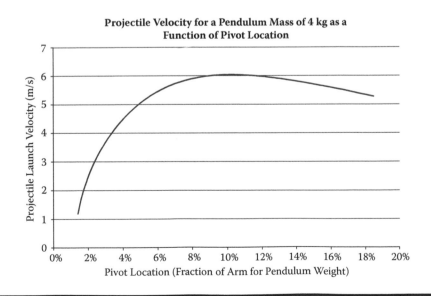

Figure 5.5 **Projectile velocity for a pendulum mass of 4 kg as a function of pivot location.**

From Figure 5.4 we see that if we place the pivot location too close to the center of gravity of the pendulum weight, or too far from it, we will need to make the pendulum weight extremely heavy (the graph increases sharply below 2.5% and above 17%). The reason is that if the pivot location is too close to the pendulum weight, the weight does not have far to fall, so there is very little potential energy in the weight when it is raised. The mass of the pendulum weight must be increased dramatically to compensate. On the other hand, if the pivot location is located too close to the projectile end of the catapult arm, then the projectile end of the arm does not travel very far, so the projectile does not absorb much kinetic energy. Again, the mass of the pendulum weight must be increased to compensate. The optimum pivot location appears from the graph to be at 10% of the total arm length. For that pivot location, a mass of 4 kg would be required to achieve the required launch velocity of the projectile. As a check on our work, we plot the projectile launch velocity as a function of the pivot location for a pendulum mass of 4 kg in Figure 5.5. The figure confirms that, for a mass of 4 kg, we need to set the pivot location to 10% of the arm length in order to achieve the required launch velocity (6 mps, in this case).

The main lesson to be gained from this is that every design exercise will involve trade-offs. Sometimes, engineering analysis and mathematics can be used to optimize these trade-offs; in other cases, physical experimentation will be required. Whether you use experimentation or engineering analysis, it is a valuable skill to be able to recognize trade-offs and discuss them with your design team. A linked house of quality, such as shown in Figure 5.3, is a systematic way to capture the discussion.

Define the Product Family

We have just completed what can be viewed as a major dive-and-surface cycle. The exploration process enabled us to generate a large number of viable integrated design concepts. This was a phase of diving into detail. The concept selection process enabled us to narrow the list of concepts to a best choice, which was a surfacing phase. We can continue diving and surfacing in this fashion with other aspects of the design (the trigger mechanism, the artistic design, and the choice of materials) until we have a fairly complete picture of our product. The one example of selecting the basic launch mechanism is sufficient for the purpose of this text to illustrate the basic process.

Rather than continue diving into detailed design, another surfacing-type activity that is appropriate at this stage is to define a product family. A *product family* is a collection of different products, designed for different market niches, that are derivative from a common product platform (Meyer and Lehnerd, 1997). All the products in the family share components and manufacturing equipment and are produced using very similar processes.

If we set up our hobbyist shop to produce weighted toy catapults, is there a range of designs that we could produce easily that would appeal to different markets? The most obvious market in our case is the age group of the grandchildren we are targeting. Thinking along these lines, is there a way to design the product and the production process so that we could easily produce catapults for different age groups? The answer is "yes" because the characteristics of the product that make it appropriate to different age groups are easily scaled.

Two design factors are critical to the safety and performance of the toy: the mass of the counterweight and the length of the catapult arm. As either of these design parameters is increased,

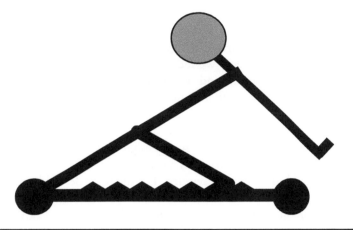

Figure 5.6   Catapult family, young child version.

the age appropriateness of the toy is also increased. We can therefore conceive of a family of toy catapults—all with the same basic design and manufacturing process, but with different combinations of counterweight and catapult arm length. With some design compromises, we could, for example, manufacture a frame that is suitable for all variations so that the frame is the common platform. The different counterweight and catapult arm combinations would be manufactured separately. The toy could thus "grow with the child." Every birthday could be celebrated by giving the child the next larger variation. Figure 5.6 and Figure 5.7 illustrate the idea of a product family for catapults.

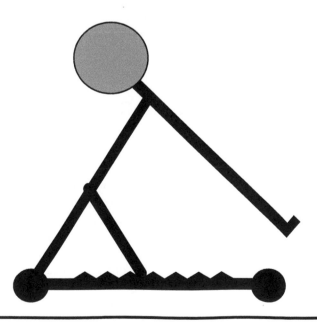

Figure 5.7   Catapult family, older child version.

Summary

Having generated a collection of integrated design concepts, we used Pugh analysis to select a design concept for further development. We identified the alternative design concepts (the columns of the concept selection matrix), and we identified the relevant product objectives (the rows), also known as product or decision attributes. In deciding which attributes were relevant, we eliminated attributes if the alternatives did not differ from one another in this attribute or if the differences were strongly correlated with another attribute that we were already considering. A simple scoring system allowed us to screen out some design concepts that were dominated by our reference concept. This permitted us to perform a more detailed analysis without wasting effort on concepts we were sure would ultimately be rejected. For the more detailed analysis, we rated the alternatives in each attribute using a five-point scale relative to the reference concept.

Finally, we used the product objective weights that we had developed earlier during our customer analysis in order to compute a score for each alternative. The alternatives, attributes, weights, ratings, and scores were presented in the form of a concept selection matrix. The team reviewed the matrix to see if the highest scoring alternative truly represented the best choice. The matrix and team decision were documented so that future teams could revisit our design decision. We were surprised to discover that our first design concept was not the best. The step of exploring the design space led us to a superior design concept.

We also considered a different type of design decision, one that differed in degree rather than in kind. We used a linked house of quality to relate subsystem engineering characteristics (design parameters) to system-level technical performance requirements. We then formulated an optimization problem to express the trade-offs involved in setting the values of the design parameters. In particular, we identified the design parameters, constraints or limits on those design parameters, and an objective function to minimize (or maximize). We chose to use engineering analysis to solve the optimization problem rather than to experiment with physical models.

Finally, we completed our exploration of the design space by considering an extension of our design to allow us to consider a range of designs based on a common product platform so that we could easily reach different market segments. This identification of a product platform is a type of surfacing activity.

Appendix: The Physics of a Catapult Design (Advanced)

In this appendix we derive the mathematics that describe the relation of design and operational parameters of the catapult to the desired performance measure. This section is not meant for the general reader. Thanks are due to Jonah Cohen for deriving the energy balance equations.

For a simple analysis, the variables of interest are displayed in Figure 5.8. We ignore friction and the mass of the catapult arm. The desired performance measure is the velocity of the projectile at the time of launch, denoted by v_2 and by $\bar{v}$, and shown in the upper right of Figure 5.8(b). The design parameters include the length of the catapult arm, l, the mass of the pendulum weight, m_1, and the position of the pivot (taken to be the fraction of the catapult arm between the pivot and the center of gravity of the pendulum weight), p. The operating parameters are those chosen at the time the catapult is used. In this case, the operating parameters would include the mass of the projectile, m_2, the starting angle above the horizontal (to which the pendulum weight is raised to supply energy to the catapult), θ, and the ending angle below the horizontal (the angle at which the projectile is launched), $\bar{\theta}$.

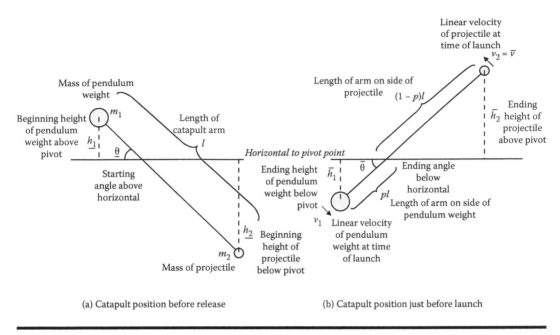

Figure 5.8   Mathematical notation for pivot location problem.

From these design and operating parameters, we derive other useful quantities, such as the length of the arm on the side of the pendulum weight, pl; the length of the arm on the side of the projectile, $(1-p)l$; the beginning height of the pendulum weight above the pivot, $h_1 = pl \sin \theta$; the beginning height of the projectile below the pivot, $h_2 = (1-p)l \sin \theta$; the ending height of the pendulum weight below the pivot, $\bar{h}_1 = pl \sin \bar{\theta}$; and the ending height of the projectile above the pivot, $\bar{h}_2 = (1-p)l \sin \bar{\theta}$. We will also be able to compute the velocity of the pendulum weight, v_1, at the time the projectile is launched. Assume that height and length are measured in meters, velocity in meters per second, mass in kilograms, and angles in radians.

We can relate all of these parameters by applying the conservation of energy law: The total potential and kinetic energy in the before-release state (Figure 5.8a) must equal the total potential and kinetic energy in the before-launch state (Figure 5.8b). Equivalently, we can equate the change in potential energy between Figure 5.8(a) and 5.8(b) with the change in kinetic energy between (a) and (b). We can use the resulting equation to create graphs such as those in Figure 5.4 and Figure 5.5.

The formula for potential energy is $P = mgh$, where P is the potential energy, measured in joules, m is the mass of the object (in kilograms), h is the height of the object (in meters), and g is the acceleration due to gravity (in meters per second-squared). The change in the potential energy of the pendulum mass from (a) to (b) is a decrease of $m_1 g (h_1 + \bar{h}_1) = m_1 g p l (\sin \theta + \sin \bar{\theta})$. To simplify, let $z = gl(\sin \theta + \sin \bar{\theta})$ because we do not intend to experiment with changes in g, l, θ, or $\bar{\theta}$. The change in potential energy of the pendulum mass is a loss of $m_1 pz$. Similarly, the change in potential energy in the projectile from Figure 5.8(a) to Figure 5.8(b) is a gain of $m_2(1-p)z$. The net decrease in potential energy from (a) to (b) is

$$\Delta P = m_1 pz - m_2(1-p)z.$$

Observe how much potential energy is released as a function of the pivot location p: Assuming that the mass of the pendulum is larger than the mass of the projectile ($m_1 > m_2$), the closer the pivot is to the projectile (i.e., as p approaches 1), the more potential energy is released.

The formula for kinetic energy is

$$K = \frac{1}{2}mv^2,$$

where K is the kinetic energy (in joules), m is the mass of the object (in kilograms), and v is the velocity of the object (in meters per second). The objects in our example are traveling in circular paths, so we want to convert linear velocity into rotational velocity. An object traveling with linear velocity v in a circular path r meters from the pivot point will complete one revolution in $2\pi r/v$ seconds (the length of the circumference divided by the velocity). The rotational velocity is therefore $v/(2\pi r)$ revolutions per second. Observe in Figure 5.8(b) that, although the pendulum weight and the projectile may have different linear velocities (v_1 and v_2, respectively), they must both have the same rotational velocity. This gives us the following relationship:

$$\frac{v_1}{2\pi pl} = \frac{v_2}{2\pi(1-p)l}.$$

Using this, we can express the linear velocity of the pendulum weight in terms of the linear velocity of the projectile and the pivot location:

$$v_1 = \frac{p}{1-p}v_2.$$

Both the pendulum and the projectile are starting from a position of rest (no kinetic energy) (Figure 5.8a). The gain in kinetic energy in going from the state in 5.8(a) to the state in 5.8(b) is the sum of the kinetic energies in Figure 5.8(b):

$$\Delta K = \frac{1}{2}m_1v_1^2 + \frac{1}{2}m_2v_2^2$$

$$= \frac{1}{2}m_1\left(\frac{p}{1-p}\right)^2 v_2^2 + \frac{1}{2}m_2v_2^2$$

By the law of conservation of energy, the net decrease in potential energy in going from the state in Figure 5.8(a) to the state in Figure 5.8(b) must equal the gain in kinetic energy. This leads to the following formula:

$$\Delta P = \Delta K,$$

which becomes

$$m_1pz - m_2(1-p)z - \frac{1}{2}m_1\left(\frac{p}{1-p}\right)^2 v_2^2 + \frac{1}{2}m_2v_2^2$$

and is rewritten as

$$2m_1 p(1-p)^2 z - 2m_2(1-p)^3 z = m_1 p^2 \bar{v}^{-2} + m_2(1-p)^2 \bar{v}^{-2}$$

What this formula makes clear is that we are not free to choose all of these design and operating parameters independently of each other. In particular, our analysis of customer requirements led us to fix the mass of the projectile, m_2, equal to 1 g and the launch velocity of the projectile, $\bar{v}$, to 6 mps. We have already grouped some variables in z that we do not intend to experiment with (the beginning and ending angles of the pendulum swing and the length of the catapult). The remaining variables are the mass of the pendulum weight, m_1, and the pivot location, p. We can rewrite the equation to make this relationship clear:

$$m_1 = m_2 \left[\frac{2z(1-p)^3 + (1-p)^2 \bar{v}^{-2}}{2zp(1-p)^2 - p^2 \bar{v}^{-2}} \right].$$

This formula was used to generate the graph in Figure 5.4. It is left as an exercise to derive the formula used to generate the graph in Figure 5.5. For any of these formulas to make sense, we must restrict the parameters to stay in the region where the denominator is positive:

$$2zp(1-p)^2 - p^2 \bar{v}^{-2} > 0.$$

This can be traced back to the need for the potential energy contribution of the pendulum weight to be greater than the kinetic energy of the pendulum weight at the time of launch.

To review, this engineering analysis made use of the following facts: trigonometric relations between angles and height, a formula for potential energy, a formula for kinetic energy, the law of conservation of energy, and a geometric conversion between linear velocity and rotational velocity. The rest of the analysis is algebraic manipulation to convert the formula into a meaningful relationship.

Discussions

1. Prior to the maiden voyage of the *Titanic,* there was great confidence that the ship was "unsinkable." Discuss how this may have influenced the lifeboat decision. If the project manager had been less susceptible to this myth, how might he have changed the hypothetical trade-off analysis matrix?
2. Discuss the role of overconfidence in design and decision making in general.
3. Consider a product family, such as the SONY Walkman™ and Apple iPod™ families, and discuss the advantages of building multiple products from common architecture elements.

Exercises

1. Concept selection: The goal of this exercise is to go through all of the steps of the concept selection process. This is a creative exercise, so you will have to contribute some of your own judgment. You are responsible for recommending a design concept for the

Table 5.12   Relative Weights for Customer Attributes of Dialysis Transport

Customer Attribute	Rounded Relative Weights (%)
Usability by an ill 95-lb person	28
Transportable up and down stairs	16
Maneuverable through doorways	16
Usability by individuals ranging in height from 4 ft, 9 in. to 6 ft, 6 in.	11
Usability with one hand free	7
Transport system of low cost	11
Battery life of dialysis unit with transport system	11

Source: Warburton, D. 2004. Getting better results in design concept selection. Medical Device & Diagnostic Industry. Adapted with permission from Canon Communications.

transportation of a portable kidney dialysis unit. A dialysis unit weights about 70 lb. The customer attributes and relative weights are given in Table 5.12. The initial set of concepts is listed in Table 5.13. Your solution should show evidence of the following steps:

a. Pick the reference concept. Explain why you picked it.
b. Screen concepts.
c. Eliminate dominated concepts.
d. Explain in simple language how the concept is dominated.
e. Rank remaining concepts.
f. Rate the remaining concepts and include a paragraph explaining ratings that might be debatable.
g. Score the remaining concepts and select the best concept.
h. Suggest a new concept (or an improvement on an existing concept).

Table 5.13   Design Concepts for Dialysis Transport

Initial Design Concepts
Backpack
Four-wheel wagon with steerable front wheels, similar to a child's wagon
Two-wheel hand truck
Suitcase
Motorized, three-wheeled cart with adjustable-height tiller

Source: Warburton, D. 2004. Getting better results in design concept selection. Medical Device & Diagnostic Industry. Adapted with permission from Canon Communications.

    i. Rate, weight, and score the new concept.

    j. Make your final recommendation of a single concept. Summarize in simple language why it is the best choice.

2. Use the concept selection methodology to pick a design concept from the list of integrated concepts you developed in the exercise from the previous chapter.

3. In simple language, describe an optimization problem involving continuous design parameters. There should be degrees of freedom, requirements to be satisfied, design parameter constraints, and an objective function to be minimized or maximized. Suggest how this might lead to a product platform approach. For example, consider the length of the flexible robot arm in the BathBot case study. What is the trade-off in making it longer or shorter?

References

Franklin, B. 1772. A letter to Joseph Priestley. As reproduced in *Mr. Franklin: A selection from his personal letters*, ed. L. W. Labaree, 1956. New Haven, CT: Yale University Press.

Meyer, M., and A. P. Lehnerd. 1997. *The power of product platforms*. New York: The Free Press.

Pugh, S. 1991. *Total design: Integrated methods for successful product engineering*. Reading, MA: Addison–Wesley.

Smith, E. D., J. S. Young, M. Piattelli-Palmarini, and A. T. Bahill. 2007. Ameliorating mental mistakes in trade-off studies. *Systems Engineering* 10 (3): 236–237.

Smith, T. 2001. Siegecraft. http:resc.net/trebuchet/trebuchet.html (accessed December 17, 2008).

Warburton, D. 2004. Getting better results in design concept selection. Medical Device & Diagnostic Industry. Canon Communications. http://www.devicelink.com/mddi/archive/04/01/006.html (accessed July 16, 2007).

Ulrich, K. T., and S. D. Eppinger. 1995. *Product design and development*. New York: McGraw–Hill.

Chapter 6

Develop the Architecture

A good solution somehow looks nice.

Robert Spinrad, 1991

The eye is a fine architect. Believe it.

Werner von Braun, 1950

Introduction

At this point, we have what we need to begin detailed design of the toy catapult. Repeated application of the explore-and-optimize cycle will lead to a complete design. Furthermore, if there were only a single designer for the catapult, rather than a team of people needing to communicate, no special documentation would be required. For larger projects, however, such as the product challenges in Appendix B, no one person will design all aspects of the system. For such projects, the overall design will be broken down into subsystems. Different individuals or teams of individuals will be responsible for the different subsystems. In this case, communication between the design teams becomes critically important. It is said that "systems tend to fail at the interfaces." In this step of the process, "develop the architecture," we take a systems view of the product or service we are developing, break it into its major subsystems, and develop a thorough understanding of how each subsystem works with the other subsystems within the context of the system to achieve the desired behavior of the overall system. The goal is to describe each subsystem with sufficient clarity that detailed design work can proceed on each subsystem independently of the design work on other subsystems.

Mars Climate Orbiter (MCO) Mishap

The following example details that a major spacecraft mishap was traced to one team assuming a computer file meant one thing and another team assuming it meant something else. This is an example of a failure at an interface.

Figure 6.1   Mars Climate Orbiter. (Source: NASA/JPL-Caltech.)

The Mars Climate Orbiter (MCO) Mission objective was to orbit Mars as the first interplanetary weather satellite and provide a communications relay for the Mars Polar Lander, which [was] due to reach Mars in December 1999. The MCO [Figure 6.1] was launched on December 11, 1998, and was lost sometime following the spacecraft's entry into Mars occultation during the Mars orbit insertion (MOI) maneuver. The spacecraft's carrier signal was last seen at approximately 09:04:52 UTC on Thursday, September 23, 1999.

The MCO Mishap Investigation Board has determined that the root cause for the loss of the MCO spacecraft was the failure to use metric units in the coding of a ground software file, "Small Forces," used in trajectory models. Specifically, thruster performance data in English units instead of metric units [were] used in the software application code titled SM_FORCES (small forces). A file called Angular Momentum Desaturation (AMD) contained the output data from the SM_FORCES software. The data in the AMD file [were] required to be in metric units per existing software interface documentation, and the trajectory modelers assumed the data [were] provided in metric units per the requirements.

[An increased frequency of momentum correction events during the 9-month flight] coupled with the fact that the angular momentum (impulse) data [were] in English, rather than metric, units, resulted in small errors being introduced in the trajectory estimate over the course of the 9-month journey. At the time of Mars insertion, the spacecraft trajectory was approximately 170 kilometers lower than planned. As a result, MCO either was destroyed in the atmosphere or re-entered heliocentric space after leaving Mars' atmosphere.

The Board recognizes that mistakes occur on spacecraft projects. However, sufficient processes are usually in place on projects to catch these mistakes before they become critical to mission success. Unfortunately for MCO, the root cause was not caught by the processes in place in the MCO project.

Mars Climate Orbiter Mishap Investigation Board Phase I Report, 1999

System Architecture and a Language for Systems

A description of the subsystems and how they work together is what we mean by the *system architecture*. Describing the architecture is difficult because design teams often speak different languages: Hardware engineers are versed in mechanical or electrical engineering and software engineers are versed in computer science. What is needed, therefore, is a common language to describe what each team needs from the other teams. This is the language of systems: functions, requirements, interfaces, components, states, etc. In this chapter, we develop a language for describing a system architecture, building on the basic concepts introduced in earlier chapters.

We continue to use the toy catapult example even though it might seem a little silly to apply this level of detail to something as simple as a toy. We do not, in fact, advance the design of the toy catapult by very much in this chapter. At the end of the chapter, the design concept is about where we left it at the end of the previous chapter. It is the explore-and-optimize cycle of the previous chapters that is most useful for advancing designs. What we achieve in this chapter, on the other hand, is a way of describing the toy catapult that would enable teams of people to work together on its design. The power of this language becomes apparent as you see how to apply it to larger and larger projects.

We use the word "system" to describe whatever it is we are designing. The system could therefore be a piece of clothing, a consumer product, a babysitting service, a piece of software, a building, an airport, a way to track aircraft in the air and prevent collisions, an online marketplace to trade used comic books, a global trade agreement, or a partially regulated economy.

Earlier, we defined a system to be a set of objects in relation to one another. Now, we mean something more specific: a *purposeful system,* which is a system that exhibits purposeful behavior. As described in the chapter on defining the problem, we find that the best way to get at the purpose of a design is to think through how it will be used. That is, the system has a set of required behaviors, called "use cases." By studying these behaviors, we can identify functions that the system must perform. The structure of the system is then designed to perform all of the functions. We suggest that the most fruitful way to approach systems design, therefore, is found in the sequence of behavior–function–structure.

We organize the development of the system architecture into a sequence of major steps:

1. Design the behavior of the system in terms of subsystem functions.
2. Design the functional control of the subsystems.
3. Design the structure of the system.

The Invention of the Post-it® Note

Of course, not all design and invention proceeds in the linear sequence of behavior–function–structure. The invention of Post-it notes by the 3M Company is an example in which function arguably came first. The story starts with the discovery by a 3M chemist, Spencer Silver, of a pressure-sensitive glue that had very weak adhesive properties. Engineers and executives at 3M knew that such glue should be useful. For many years, they searched for a marketable application of the glue, but without success. This, in a way, is "function searching for purpose." However, even in this example, we see that, ultimately, the motivation for the invention itself came from a use case. Art Fry, a chemical engineer for 3M, was frustrated that bookmarks kept falling out of his choir books at church (use case behavior). He reasoned that he wanted a bookmark that would not fall out but was also easily removable (functional requirements). He suddenly remembered Silver's discovery and used it to design the first sticky-note (structure).

Design the Behavior

Even though one typically thinks of design in terms of the structure of the final system, we place a greater emphasis on the behavior of the system by putting it first. We find that thinking about the system in terms of its behaviors has great integrative power. Functional analysis, control, structure, and verification all flow naturally from a consideration of use case behaviors first. In this section, we start from the use case behaviors defined in Chapter 2 and use these to establish functional requirements for each of the subsystems.

We organize the development of an architecture into the following steps:

1. Review the system use cases, context, and requirements.
2. Map the behaviors to subsystems.
3. Identify messages, triggers, and interfaces.
4. Identify system states.
5. Set targets for behavior.
6. Extract functional requirements.
7. Trace derived requirements to originating requirements.

Review Use Cases, Context, and Functional Requirements

We begin the design of behavior by reviewing the results of our previous analysis. In Chapter 2 and in Appendix C, we conducted a use case behavioral analysis of the toy catapult. Table 6.1 lists the use case behaviors, both primary and secondary, that we considered. Recall that we attached priorities to these use cases to manage the effort spent on developing them. High-priority use cases (priority "H") were developed in detail. Lower priority use cases (priorities "M" and "L") would be developed in much less detail or ignored.

We reiterate the importance of developing secondary use case behaviors. It is now understood that much of the architecture of the final product comes from the need to satisfy these secondary concerns. For example, Bass, Clements, and Kazman (2003) emphasize that all software would be written as a single block of code if it were not for the secondary concerns of reusability and maintainability. It is these secondary concerns that have led to modular software architectures and the modern paradigm of object-oriented programming.

For each use case behavior, we developed a detailed thread from stimulus to response of how the behavior would play out among the user, the system, and other external entities. Because of their length and number, we do not reproduce the threads here.

The behavioral analysis of use cases yielded a context matrix (Table 6.2) for the toy catapult showing the system (the toy catapult, in this example) in relation to the external entities with which it must interact. The context matrix defines the system boundary and the major interfaces between the designed system and its external entities. The context matrix is designed to be the "big picture" view of our system, a view that fits on a single sheet of paper.

The analysis of Chapter 2 also yielded a set of system-level functional requirements. Table 6.3 lists those requirements. These are referred to as "originating requirements" to distinguish them from the requirements we will derive in this section.

Table 6.1   Prioritized Use Case Behaviors

Use Case Behaviors	Priority
Child chooses toy	M
Child retrieves toy	L
Child plays with toy	H
Parent teaches child	L
Parent puts toy away	L
Parent entertains child with train and toy	H
Parent entertains child with train and two copies of toy	H
Secondary use case behaviors	Priority
Child releases armed toy near face of self or of another child	H
Child aims projectile at eyes of self or of another child	H
Child uses pet rodent as projectile	H
Child plays with toy repeatedly	H
Child drops or throws toy	H
Child leaves toy outside in bad weather	M

Note that the functional requirements are derived as logically necessary to fulfill the design mission and to enable the use case behaviors. They are not, by themselves, sufficient. We can design a toy catapult that satisfies all of these requirements, but fails to delight the parent and child. That is, the requirements do not state how well the functions are to be performed. The toy may even fail to perform its primary use case behavior: having a child arm, load, trigger, and launch a projectile. That is, these are isolated functional requirements. Suppose, for example, that the act of triggering the toy catapult always knocks the projectile out of the receptacle. That particular circumstance does not violate any of the functional requirements (unless we reword them to cover this circumstance), but it will result in a completely unsatisfactory design. Ultimately, it is the behavior itself—everything working together—that is important.

In Chapter 3, using a house of quality analysis, we developed specific targets for technical performance measures. However, none of these particular targets affect our description of the behavior of the system, so we defer their consideration to a later section.

Finally, recall the results of our exploration phase in Chapter 4. There, we assimilated the results of our concept generation activity into an abstract list of the primary subsystems of the toy catapult (Table 6.4). Recall that the process of identifying these subsystems came from a dive-and-surface approach in which functional decomposition played a critical role. We analyzed the behavior into detailed functions and then used the affinity process to summarize back up to a subsystem description level.

This completes our review of the major outputs from previous steps in the design process.

Table 6.2 Context Matrix for Toy Catapult

Is Related to	Child	Parent	Toy	Projectile	Other Moving Toys	Small Pets	Play Surfaces
Child			retrieves and plays with repeatedly; arms and loads; triggers release of	places in toy receptacle	anticipates triggering event from	inappropriately uses as projectile	drops or throws toy against
Parent	teaches, entertains, and trains in safety procedures		stores; positions for chain reaction behavior		aligns trajectory with toy release mechanism		
Toy	amuses but does not harm	warns of dangers and suggests appropriate age of child		holds and launches		does not harm	survives impact with
Projectile	amuses but does not harm		triggers copy 2 of				
Other moving toys	amuses		trigger				
Small pets			escape from or survive launch from				
Play surfaces							

Note: Thick border encloses system boundary and major interfaces.

Table 6.3   Originating Functional Requirements for Toy Catapult

Index	Originating Requirements	Abstract Function Name
OR.1	The system shall detect the command to arm.	Detect
OR.2	The system shall secure its receptacle in the armed position. The force required to arm the system may be supplied by the system (e.g., battery power) or by the child.	Arm
OR.3	The system shall hold the projectile in its receptacle.	Load
OR.4	The system shall detect the command to release from a passing toy (lateral direction) and also from a child or falling projectile (vertical direction).	Trigger
OR.5	The system shall be capable of ejecting lightweight balls and dolls from its receptacle.	Eject
OR.6	The system shall notify the parent of the dangers and inherent risks of the system.	Warn
OR.7	The system shall suggest to the parent the appropriate age of the child to use the system.	Suggest age
OR.8	The moving parts of the system shall be prevented from striking the face of a child with sufficient force to cause a bruise.	Protect face
OR.9	The system shall not launch projectiles from the child's domain with sufficient force to cause damage to the child or the projectile.	Constrain launch force
OR.10	The system shall enable the escape of a pet rodent from the receptacle.	Enable pet escape
OR.11	The system shall successfully complete each cycle of use for many cycles.	Repeat
OR.12	The system shall survive a collision with a hard surface and remain in working order.	Survive collision
OR.13	Verification of all requirements shall be conducted in such a way as to preserve the health of human and animal subjects.	Test safely

Table 6.4   Toy Catapult Subsystems

Toy				
Energy storage system	Locking system	Projectile containment system	Trigger system	Projectile launch system

Table 6.5 Operational Description Template for "Child Plays with Toy"

Child Plays with Toy
Initial conditions
1. System is in the unloaded state

Map Behaviors to Subsystems

The major activity in the process of designing behavior is to take the use case behaviors and map them into the subsystems we have identified. We are not starting from scratch. We have all the detailed threads that were created during the initial behavioral analysis phase. In this step, we revisit each of those threads and consider them in more detail. It is also an opportunity to correct and modify the threads based on our current understanding of the design problem.

To describe behavior at the subsystem level, we will take the spreadsheet template used in Chapter 2 and extend it in several valuable ways. The resulting template is called an *operational description template* (ODT). Len Karas and Donna Rhodes of Lockheed Martin developed and popularized the technique (1987, 1993, 1999 as well as Karas 2001, 2006).

The operational description template describes a single thread of behavior from initial stimulus through to final response. The template begins with a description of the use case and the initial conditions (Table 6.5).

The next section of the template (Table 6.6) is a tabular view of the thread, using columns to represent the internal and external entities of the system. The external entities are taken from the context matrix (Table 6.2). The user (who often provides the stimulus) is always placed in the left-most column. All other external entities are placed in the rightmost columns. The internal entities are the subsystems (Table 6.4). The rows of the table refer to the sequence of events or activities. Each row contains a single event or activity unless the activities in the row truly take place concurrently. The activities are stated as functional requirements ("the system shall…"). The sequence can be read from top to bottom, jumping back and forth between the columns. So far, the ODT looks very much like the spreadsheet template used in Chapter 2. What is different is that the functions that the system performs are allocated to the different subsystems. This is our first view of how the different subsystems must work together to achieve a unified behavior.

There is more detail here than what we sketched during our initial behavioral analysis in Chapter 2. In particular, we noticed this time through that there was nothing to mark the transition from the child pushing the receptacle into position to the next activity of the child loading the receptacle. We realized that it would be good to recognize that the child stops pushing at some point and that there is an interaction with the energy storage system when this happens. We decided to include a function "the system shall limit the energy it accepts" in the energy storage system column and an event, "the child stops pushing," in the operator column.

What level of detail is important for this analysis? If an activity belongs solely to a single column (a single subsystem), then it is not necessary to analyze it further. When an activity involves multiple subsystems, then it must be analyzed into its constituent parts to see what each subsystem contributes to the overall functionality.

The final section of the operational description template (Table 6.7) includes a description of the proper ending conditions of the behavior and any notes that help to explain the behavior. Additional features in the operational description template will be developed in the next subsections. These include ways to represent messages, interfaces, states, and targets.

Table 6.6 Operational Description Template for "Child Plays with Toy"

Operator (Child)	Toy					
	Energy Storage System	Locking System	Projectile Containment System	Trigger System	Projectile Launch System	Projectile
The child pushes the receptacle						
	The system shall store or accept external energy					
The child continues pushing until the receptacle is in position **Yes** No						
		The system shall detect the receptacle in proper position				
		The system shall secure its receptacle				
				The system shall enter the armed state		

(Continued)

Table 6.6 Operational Description Template for "Child Plays with Toy" (Continued)

Child	Energy Storage	Locking	Containment	Trigger	Launch	Projectile
	The system shall limit the energy it accepts					
The child stops pushing						
The child loads the receptacle with projectile						
			The system shall accept and hold the projectile in its receptacle			
						The projectile sits in the receptacle
The child triggers the release (see note 1)						
				The system shall detect the command to release (laterally or vertically) from the child		

					The projectile flies through the air and lands some distance away
		The system shall apply translational energy to receptacle	The system shall stop and enter the unloaded state		
				The system shall release the contents of the receptacle	
The system shall release the receptacle					
	The system shall convert energy to translational energy				

Table 6.7 Operational Description Template for "Child Plays with Toy"

Ending Conditions	
1.	The toy is in the unloaded state

Notes:

1. This behavior must be repeated with different stimuli (child triggers release, passing toy train triggers release, falling toy triggers release)

We show only one operational description template for the toy catapult example—namely, the "child plays with toy" case. The complete set of use case behaviors documented with ODTs can be found in the downloadable spreadsheet that accompanies this section.

Identify Messages, Triggers, and Interfaces

Once we have allocated the functionality of the system to different subsystems, it is clear that something must be happening to trigger and coordinate these different functions. That is, an event or activity in one column must somehow activate an activity or trigger an event in another column. We insert a special row into the operational description template whenever we want to describe some communication that is taking place. Give this row a different background color (we have used gray) and identify the event or message in the cell of this row under the column where the event or message originates. We refer to these colored rows as *interface rows:* They represent an interface between two or more subsystems and/or external entities. Table 6.8 and its continuations in Table 6.9 and Table 6.10 illustrate for the toy catapult.

For example, the connection between the child pushing the receptacle and the system detecting the receptacle in the proper position is an event—namely, the event that the receptacle truly is in position. By isolating that event on a row of its own, we can begin to think generically of different ways in which the information of this event can be transferred to the locking system.

In purely mechanical systems, such as the toy catapult, it can be difficult at first to think in terms of messages and transfers. Consequently, some of the connections may seem a little artificial. For example, "the system shall accept and hold the projectile in its receptacle" is most likely a passive activity, so it is difficult to think of it being "activated." Nevertheless, we want to indicate some connection between the child's action of loading the receptacle and the new state of the system ("receptacle loaded") so we invent the concept of the "load event" signified by the successful transfer of material to the receptacle.

In electromechanical systems or software systems, these interfaces are usually much easier to identify: They will correspond to signals, messages, or software function calls. For example, imagine an automated checkout counter at a grocery store: The system can detect when you load an item into the grocery bag from the weight of the bag on the scales. In fact, it will not let you scan the barcode of a new item until it detects that the previously scanned item (or something with the identical weight) has been placed in the bag. Thus, the "load event" in the grocery checkout system is likely an important signal sent from the check-weight subsystem to the central processor subsystem.

Identify System States

Another useful way to think of the behavior of a system is to identify the different states in which the system can be found, and to note the transitions from one state to another. We do this in the operational description template by adding a column to describe the system state. The *state of the system* is a description of the system that permits you to distinguish how the behavior has changed the system. For

Table 6.8 Operational Description Template with Interface Rows

Operator (Child)	Energy Storage System	Locking System	Projectile Containment System	Trigger System	Projectile Launch System	Projectile
			Toy			
The child pushes the receptacle						
Energy transfer ("energy in")						
	The system shall store or accept external energy					
The child continues pushing until the receptacle is in position **Yes** No						
Information event ("receptacle in position")						
		The system shall detect the receptacle in proper position				
		The system shall secure its receptacle				
		Information event ("receptacle secured")				
				The system shall enter the armed state		

Table 6.9 Operational Description Template with Interface Rows

| Child | Toy | | | | | Projectile |
	Energy Storage	Locking	Containment	Trigger	Launch	
	The system shall limit the energy it accepts			The system shall enter the armed state		
	Information event ("maximum energy reached")					
The child stops pushing						
The child loads the receptacle with projectile(s)						
Material transfer ("load")			The system shall accept and hold the projectile in its receptacle			
						The projectile sits in the receptacle

Table 6.10 Operational Description Template with Interface Rows

Child	Energy Storage	Locking	Containment	Trigger	Launch	Projectile
The child triggers the release (see note 1)						
Information event ("trigger")						
				The system shall detect the command to release (laterally or vertically) from the child		
				Information event ("release command")		
		The system shall release the receptacle				
		Information event ("release action")				
	The system shall convert energy to translational energy					

(Continued)

Table 6.10 Operational Description Template with Interface Rows (Continued)

Child	Energy Storage	Locking	Containment	Trigger	Launch	Projectile
	Energy transfer ("energy out")					
					The system shall apply translational energy to receptacle	
					Information event ("stop")	
					The system shall stop and enter the unloaded state	
					Information event ("release contents")	
			The system shall release the contents of the receptacle			
			Material transfer ("launch")			
						The projectile flies through the air and lands some distance away

Toy (spanning Energy Storage, Locking, Containment, Trigger, Launch)

example, how is the toy catapult different after the receptacle has been locked into position than before? We say it has entered the "armed state." Because we have separated the locking system from the trigger system, we acknowledge that "locked" and "armed" are two different aspects of the state. "Locked" refers to the receptacle being secured. "Armed" refers to the trigger mechanism being ready to detect the command to be fired. They describe the states of two different subsystems. Table 6.11 and Table 6.12 show the operational description template with a system state column added.

This analysis forced us to recognize that there were periods of the behavior when the toy catapult was unstable. As the child is pushing on the receptacle, the energy storage system is unstable until the locking system has detected the receptacle in position and secured it. If the child were to release the receptacle prematurely, there might be a risk of injury. This line of thinking caused us to expand the list of secondary use case behaviors to explore what happens in this event. That is why the activity is labeled as "the child continues pushing until the receptacle is in position: yes, no" with the "yes" branch highlighted. We created a separate operational description template (not shown) to handle the behavior "child fails to arm toy." In that behavior thread, the same activity "the child continues pushing until the receptacle is in position: yes, no" appears, but this time with the "no" branch highlighted. That behavior describes a behavior in which the toy catapult safely releases the energy from the energy storage system. The ODT for this behavior can be found in the accompanying downloadable spreadsheet.

Set Targets for Behavior

The operational description template can be used to capture other information. For example, if there are timing restrictions or targets, we can include a column to represent the planned duration of an activity or sequence of activities. Table 6.13 and Table 6.14 extend the toy catapult behavior to include targets for the time to arm and the time to release. For the toy catapult, these targets are not actually critical to the design. They are included simply to show how this type of information can be included in an operational description template.

Extract Functional Requirements

Creating the operational description templates for a system can be a significant amount of work. It is useful to get the entire project team involved in this activity. That way, each team member thinks in terms of the overall behavior of the system before diving into subsystem level detail. Make each team member responsible for generating ODTs for several of the use case behaviors. A good rule of thumb is 10 ODTs per team member.

After the behavior of the system has been thoroughly mapped to the subsystems, collect all of the ODTs together into one place (e.g., multiple sheets in the same Excel workbook). Now, go through each ODT and copy the functional requirements for each subsystem to a separate sheet. Table 6.15 shows the result of collecting all of the functional requirements for each subsystem from the several ODTs we created for the toy catapult. Karas and Rhodes refer to this process as "*walking the columns*" because, for each subsystem, we collect its functional requirements by simply going down the corresponding column in each ODT and capturing every new functional requirement that we encounter. The result is an *object-oriented view* of the system: The system is a collection of objects and each object is responsible for certain functions.

Observe that we can now begin to think of making different teams responsible for different subsystems. We have described the bundle of functional requirements that must be satisfied by each subsystem for the entire system to behave as desired. Different teams can focus on these different bundles of requirements and proceed to design solutions to those bundles.

Table 6.11 Operational Description Template with System States

Operator (Child)	Toy						System State
	Energy Storage System	*Locking System*	*Projectile Containment System*	*Trigger System*	*Projectile Launch System*	*Projectile*	*System State*
The child pushes the receptacle							Unloaded, stable, unarmed
Energy transfer ("energy in")	The system shall store or accept external energy						Unloaded, unstable, unarmed
The child continues pushing until the receptacle is in position **Yes** No							
Information event ("receptacle in position")		The system shall detect the receptacle in proper position					
		The system shall secure its receptacle					Unloaded, locked, unarmed

		Information event ("receptacle secured")								
				The system shall enter the armed state					Unloaded, locked, armed	
The child stops pushing										
The child loads the receptacle with projectile(s)										
Material transfer ("load")							The system shall accept and hold the projectile in its receptacle		Loaded, locked, armed	
	The system shall limit the energy it accepts									The projectile sits in the receptacle
	Information event ("maximum energy reached")									

Table 6.12 Operational Description Template with System States

Ch-Id	Toy						System State
	Energy Storage	Locking	Containment	Trigger	Launch	Projectile	
The child triggers the release (see note 1)							
Information event ("trigger")							
				The system shall detect the command to release (laterally or vertically) from the child			
				Information event ("release command")			
		The system shall release the receptacle					Loaded, released (unstable), unarmed
		Information event ("release action")					
	The system shall convert energy to translational energy						

			Loaded, stable, unarmed			Unloaded, stable, unarmed		
								The projectile flies through the air and lands some distance away
		The system shall apply translational energy to receptacle	Information event ("stop")	The system shall stop	Information event ("release contents")			
					The system shall release the contents of the receptacle	Material transfer ("launch")		
Energy transfer ("energy out")								

Table 6.13 Operational Description Template with Targets

Child	Toy							
	Energy Storage	Locking	Containment	Trigger	Launch	Projectile	State	Timing
The child pushes the receptacle							Unloaded, stable, unarmed	
Energy transfer ("energy in")	The system shall store or accept external energy						Unloaded, unstable, unarmed	
The child continues pushing until the receptacle is in position **Yes** No								
Information event ("receptacle in position")		The system shall detect the receptacle in proper position						
		The system shall secure its receptacle					Unloaded, locked, unarmed	1 second

	Unloaded, locked, armed			The system shall enter the armed state				Loaded, locked, armed	
Information event ("receptacle secured")									The projectile sits in the receptacle
The system shall limit the energy it accepts									
Information event ("maximum energy reached")									
The child stops pushing									
The child loads the receptacle with projectile(s)					The system shall accept and hold the projectile in its receptacle				
Material transfer ("load")									

Table 6.14 Operational Description Template with Targets

Child	Toy						State	Timing
	Energy Storage	Locking	Containment	Trigger	Launch	Projectile		
The child triggers the release (see note 1)								
Information event ("trigger")								
				The system shall detect the command to release (laterally or vertically) from the child				
				Information event ("release command")				
		The system shall release the receptacle						
		Information event ("release action")						
							Loaded, released (unstable), unarmed	
	The system shall convert energy to translational energy							
								1 second

				Loaded, stable, unarmed		Unloaded, stable, unarmed		
								The projectile flies through the air and lands some distance away
	The system shall apply translational energy to receptacle	Information event ("stop")	The system shall stop	Information event ("release contents")				
					The system shall release the contents of the receptacle	Material transfer ("launch")		
Energy transfer ("energy out")								

Table 6.15   Object-Oriented View: Functional Requirements by Subsystem

Toy				
Energy Storage System	*Locking System*	*Projectile Containment System*	*Trigger System*	*Projectile Launch System*
The system shall store or accept external energy	The system shall detect the receptacle in proper position	The system shall accept and hold the projectile in its receptacle	The system shall enter the armed state	The system shall apply translational energy to receptacle
The system shall limit the energy it accepts	The system shall secure its receptacle	The system shall release the contents of the receptacle	The system shall detect the command to release (laterally or vertically) from the child	The system shall stop
The system shall convert energy to translational energy	The system shall release the receptacle			The moving parts of the system shall be prevented from striking the face of a child with sufficient force to cause a bruise
The system shall release stored energy				The system shall not launch a blunt projectile with sufficient force to penetrate a child's eye
				The system shall not launch a rodent with sufficient force to injure the rodent when it lands

Trace-Derived Requirements to Originating Requirements

From studying behavior in detail, we come to an understanding of function. Our list of functional requirements has grown. Table 6.16 lists the functional requirements derived from the operational description templates. For the toy catapult, we have only 18 functional requirements. For large systems, the number will be in the thousands.

It is critically important to be able to take any detailed requirement and trace it back to its origin. This is because you will often find requirements to be in conflict with each other. It may be impossible to design a subsystem to satisfy all of the requirements imposed upon it. In that

Table 6.16 Derived Functional Requirements

Index	Derived Functional Requirement	Function Name
DR.1	The system shall store or accept external energy.	Store energy
DR.2	The system shall limit the energy it accepts.	Limit energy
DR.3	The system shall convert energy to translational energy.	Convert energy
DR.4	The system shall release stored energy.	Release energy
DR.5	The system shall detect the receptacle in proper position.	Detect arming position
DR.6	The system shall secure its receptacle.	Secure receptacle
DR.7	The system shall release the receptacle.	Release receptacle
DR.8	The system shall accept and hold the projectile in its receptacle.	Hold projectile
DR.9	The system shall release the contents of the receptacle.	Release projectile
DR.10	The system shall enter the armed state.	Enter armed state
DR.11	The system shall detect the command to release (laterally or vertically) from the child.	Detect release command
DR.12	The system shall apply translational energy to receptacle.	Apply energy
DR.13	The system shall stop.	Stop
DR.14	The moving parts of the system shall be prevented from striking the face of a child with sufficient force to cause a bruise.	Do not hurt child's face
DR.15	The system shall not launch a blunt projectile with sufficient force to penetrate a child's eye.	Do not hurt child's eye
DR.16	The system shall not launch a rodent with sufficient force to injure the rodent when it lands.	Do not injure rodent
DR.17	The system shall successfully complete each cycle of use.	Survive repetition
DR.18	The system shall survive a collision with a hard surface in working order.	Survive collision

case, the requirements will have to be changed. Changing a requirement without knowing where it came from is a prescription for failure, like saying we will eliminate the wheels from the vehicle because we have exceeded the weight restriction. In the case of functional requirements, we should be able to take any *derived functional requirement* and identify which originating requirement it came from. Table 6.17 does that for the derived requirements of Table 6.16 by mapping them to

Table 6.17 Trace to Originating Requirements

Index	Derived Functional Requirement	Function Name	Derived From
DR.1	The system shall store or accept external energy.	Store energy	OR.5
DR.2	The system shall limit the energy it accepts.	Limit energy	OR.9
DR.3	The system shall convert energy to translational energy.	Convert energy	OR.5
DR.4	The system shall release stored energy.	Release energy	OR.5
DR.5	The system shall detect the receptacle in proper position.	Detect arming position	OR.1
DR.6	The system shall secure its receptacle.	Secure receptacle	OR.2
DR.7	The system shall release the receptacle.	Release receptacle	OR.5
DR.8	The system shall accept and hold the projectile in its receptacle.	Hold projectile	OR.3
DR.9	The system shall release the contents of the receptacle.	Release projectile	OR.5
DR.10	The system shall enter the armed state.	Enter armed state	OR.2
DR.11	The system shall detect the command to release (laterally or vertically) from the child.	Detect release command	OR.4
DR.12	The system shall apply translational energy to receptacle.	Apply energy	OR.5
DR.13	The system shall stop.	Stop	OR.5
DR.14	The moving parts of the system shall be prevented from striking the face of a child with sufficient force to cause a bruise.	Do not hurt child's face	OR.8
DR.15	The system shall not launch a blunt projectile with sufficient force to penetrate a child's eye.	Do not hurt child's eye	OR.9
DR.16	The system shall not launch a rodent with sufficient force to injure the rodent when it lands.	Do not injure rodent	OR.9
DR.17	The system shall successfully complete each cycle of use.	Survive repetition	OR.11
DR.18	The system shall survive a collision with a hard surface in working order.	Survive collision	OR.12

the requirement numbers in Table 6.3 (the "derived from" column). Such a table is a major tool used in managing requirements' traceability.

Table 6.18 takes the information from Table 6.17 and Table 6.3 and presents it in the form of a *requirements trace matrix,* in which we list derived requirements in one direction (rows, in this

Table 6.18 The Requirements Trace Matrix

(row) is derived from (column)		OR.1 The system shall detect the command to arm.	OR.2 The system shall secure its receptacle in the armed position. The force required to arm the system may be supplied by the system (e.g., battery power) or by the child.	OR.3 The system shall hold the projectile in its receptacle.	OR.4 The system shall detect the command to release from a passing toy (lateral direction) and also from a child or falling projectile (vertical direction).	OR.5 The system shall be capable of ejecting lightweight balls and dolls from its receptacle.	OR.6 The system shall notify the parent of the dangers and inherent risks of the system.	OR.7 The system shall suggest to the parent the appropriate age of the child to use the system.	OR.8 The moving parts of the system shall be prevented from striking the face of a child with sufficient force to cause a bruise.	OR.9 The system shall not launch projectiles from the child's domain with sufficient force to cause damage to the child or the projectile.	OR.10 The system shall enable the escape of a pet rodent from the receptacle.	OR.11 The system shall successfully complete each cycle of use for many cycles.	OR.12 The system shall survive a collision with a hard surface and remain in working order.	OR.13 Verification of all requirements shall be conducted in such a way as to preserve the health of human and animal subjects.
The system shall store or accept external energy.	DR.1					X								
The system shall limit the energy it accepts.	DR.2									X				
The system shall convert energy to translational energy.	DR.3					X								
The system shall release stored energy.	DR.4					X								
The system shall detect the receptacle in proper position.	DR.5	X												
The system shall secure its receptacle.	DR.6		X											
The system shall release the receptacle.	DR.7					X								

(Continued)

Table 6.18   The Requirements Trace Matrix (Continued)

The system shall accept and hold the projectile in its receptacle.	DR.8			X												
The system shall release the contents of the receptacle.	DR.9					X										
The system shall enter the armed state.	DR.10		X													
The system shall detect the command to release (laterally or vertically) from the child.	DR.11				X											
The system shall apply translational energy to receptacle.	DR.12					X										
The system shall stop.	DR.13					X										
The moving parts of the system shall be prevented from striking the face of a child with sufficient force to cause a bruise.	DR.14								X							
The system shall not launch a blunt projectile with sufficient force to penetrate a child's eye.	DR.15										X					
The system shall not launch a rodent with sufficient force to injure the rodent when it lands.	DR.16										X					
The system shall successfully complete each cycle of use.	DR.17												X			
The system shall survive a collision with a hard surface in working order.	DR.18													X		

case) and originating requirements in another direction (columns, in this case) and place a mark in each cell for which a derived requirement can be traced to an originating requirement. We see immediately from this matrix that the originating requirement OR.5 ("the system shall be capable of ejecting lightweight balls and dolls from its receptacle") has given rise to the most derived requirements. OR.5 is one of the primary originating requirements. Several originating requirements have no derived requirements (OR.6, OR.7, OR.11, and OR.14). A systems architect will suspect that the analysis is not complete if no derived requirements have been discovered.

The Centrality of the Operational Description Template

We pause at this point to reflect on the centrality of the operational description template in describing a system architecture. Figure 6.2 shows a graphic of an ODT and overlays the types of

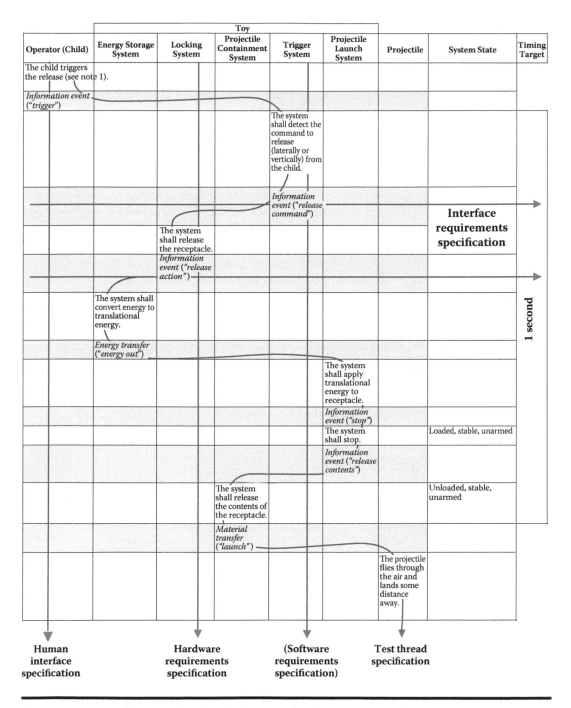

Figure 6.2 How the operation description template integrates specifications.

information that originate at the ODT level. The leftmost column describes what an operator does with the system. It is the starting point for establishing the *human interface specification*. For example, someone who is documenting a user's manual could study this column to find what user procedures need to be explained. As described in the previous section (see Table 6.15), the subsystem columns provide the functional and interface specifications for individual subsystems. Some of the columns may represent hardware subsystems; other columns may represent software subsystems. Thus, by walking the columns, we can, in general, extract *hardware requirements specifications* and *software requirements specifications*. The behavior thread itself will be the basis for test procedures, so the ODT is also the primary source for developing *test thread specifications*. Finally, the interface rows are the primary source for developing *interface requirements specifications*.

As mentioned, the toy catapult system has no software subsystems; however, in a more complex system, such as a rocket launcher, some subsystems, such as the trigger subsystem, would be largely software controlled. Because the ODT is functionally oriented, it can show hardware and software subsystems participating in the use case behavior without the need to distinguish which is which. In fact, the decision as to which subsystems to handle by mechanical means and which to handle by software means can be deferred to a later point in the design cycle.

As we proceed to develop a description of the system architecture, we will be looking at interfaces, functional requirements, and threads in a more isolated fashion. Figure 6.2 is valuable for expressing the essential unity of these ideas.

You can use operational description templates to capture more information than what we have shown here. For example, in a software system, the interface rows may represent function calls or message transfers. You could extend the template to include columns where you describe the function call arguments or message information content in each interface row. We show later how test procedure information can be layered into the ODT as well.

Design the Flow and Control

It is easy to get lost in the detail of the behavioral descriptions. In this section, we describe two ways of summarizing the behavior of any system: a functional view and a state change view. Describing behavior is a diving activity; creating functional and state-change views is a surfacing activity.

Identify Functional Relationships

Behaviors can be seen as functions in different subsystems interacting with each other. Therefore, mapping the relationships between functions can provide an aggregate view of behavior. A *functional view* of a system is simply a description of how the functions are related to each other in creating behaviors. The easiest way to create a functional view is to create a *functional interrelationship matrix* in which both the rows and columns list functions (in the same order) and the cells of the table describe known relationships.

To illustrate, Table 6.19 displays a functional interrelationship matrix formed from the derived functional requirements of Table 6.16. Note that we now use the abstract function names to save space. We review all of the behavioral descriptions to discover the ways in which these functions are related. Typically, the relation is of the form "function A triggers function B." The word "trigger" here is not to be confused with the trigger mechanism of the toy catapult. It simply indicates causation. In a software application, we might replace the word "trigger" with "calls"; in an electromechanical system, we might replace it with "activates" or "actuates." Note that we can easily

Table 6.19 Functional Interrelationships

(Row) Acts on or Influences (Column)	Store Energy	Limit Energy	Convert Energy	Release Energy	Detect Arming Position	Secure Receptacle	Release Receptacle	Hold Projectile	Release Projectile	Enter Armed State	Detect Release Command	Apply Energy	Stop
Store energy		If maximum stored energy is reached, triggers		If child prematurely releases, triggers	If child continues to push receptacle into position, triggers								
Limit energy													
Convert energy												triggers	
Release energy													
Detect arming position						triggers							
Secure receptacle										triggers			
Release receptacle			triggers										
Hold projectile									precedes				
Release projectile													
Enter armed state											precedes		
Detect release command							triggers						
Apply energy													triggers
Stop									triggers				

represent conditional triggers through the use of "if" clauses, as in "if maximum stored energy is reached, then function 'store energy' triggers function 'limit energy.'"

This process can also reveal logical relationships, in addition to causation. For example, we realize that it is impossible to perform the function "release projectile" if the function "hold projectile" failed or did not happen. We use the phrase "function A precedes function B" to capture such logical relationships. We have highlighted these relationships by using italics.

Observe how this matrix summarizes many pages of behavioral description. Functional views tend to be very compact ways of viewing behavior. Analysis of functional relationships can improve your understanding of the system. For example, by creating this matrix, we discovered the implicit requirement that "enter armed state" must precede "detect release command."

A functional interrelationship matrix such as Table 6.19 is an example of a *precedence matrix*, which is a square matrix in which the rows and columns represent activities, functions, or events that occur in time. The matrix has an entry in a cell if the row activity, function, or event logically or causally precedes the column activity, function, or event. In general, a precedence matrix is easier to read if the rows and columns are sorted so that most of the entries lie above the diagonal of the matrix (spreadsheet skill: reordering the rows and columns of a matrix). In Table 6.20 we have reordered the rows (and columns) so that all of the entries are above the main diagonal (the shaded cells).

In some cases, it may be impossible to reorder the rows and columns of a precedence matrix so that all entries are above the diagonal. If that is the case, it indicates that the behavior has a loop. Loops must be studied in detail to see if they will result in undesirable behaviors. There are two main types of undesirable behavior in systems: gridlock and infinite loops. *Gridlock* occurs when there is a cycle of precedence relations so that no activity can take place (e.g., when four drivers enter an intersection and each driver wants to make a left-hand turn: Car 1 cannot turn until car 2 turns, car 2 cannot turn until car 3 turns, and so on until car 4 cannot turn until car 1 turns). An *infinite loop* occurs when activities in a cycle trigger one another without ceasing. Infinite loops can be undesirable when some system input (e.g., computer memory) is consumed without limit (until the program crashes) or some output (e.g., sound) is amplified without limit (until the speakers are destroyed).

We have been emphasizing tabular presentations of interrelationships. Most of these presentations have diagrammatic equivalents. For example, the diagrammatic equivalent of the functional interrelationship matrix is the functional flow block diagram. A *functional flow block diagram* depicts functions as rectangles ("function blocks") and uses rectilinear arrows to represent triggering relations. Dashed rectilinear arrows are used to represent precedence relations. Figure 6.3 is a functional flow block diagram for the toy catapult. It is important to recognize that Table 6.20 and Figure 6.3 are equivalent. There is no information in one that is not conveyed in the other. You will likely prefer the diagram for its visual appeal, but the matrix is easier to create.

Summarize State Changes

Another way to describe behavior is by using the *state change view*, which focuses on the different states that the system can be in during its many behaviors and describes the events that transform the system from one state to another. An *event* is a single epoch in time when something of significance occurs. An easy way to present a state change view is with a *state change matrix*, in which the rows and columns are the different states and the cells describe events that transform the system from one state to another.

Table 6.20 Reordered Functional Interrelationships

(Row) Acts on or Influences (Column)	Store Energy	Limit Energy	Release Energy	Detect Arming Position	Secure Receptacle	Hold Projectile	Enter Armed State	Detect Release Command	Release Receptacle	Convert Energy	Apply Energy	Stop	Release Projectile
Store energy		If maximum stored energy is reached, triggers	If child prematurely releases, triggers	If child continues to push receptacle into position, triggers									
Limit energy													
Release energy													
Detect arming position					triggers								
Secure receptacle							triggers						precedes
Hold projectile													
Enter armed state								precedes					
Detect release command									precedes				
Release receptacle										triggers			
Convert energy											triggers		
Apply energy												triggers	
Stop													triggers
Release projectile													

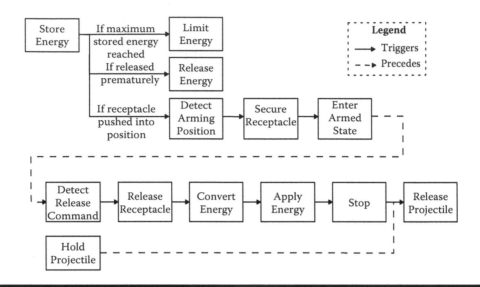

Figure 6.3 Functional flow block diagram for toy catapult.

Table 6.21 is a state change matrix for the toy catapult. It was constructed by reviewing all of the operational description templates and collecting the various states that were identified. This list forms the row and column headings of the matrix. We then give names to the events that transform the states and place them in the corresponding cells. As an example, we can read from the figure that "the event 'receptacle in position' marks the transition from the state 'unloaded, unstable, unarmed' to the state 'unloaded, locked, unarmed.'"

As with the analysis of functional relationships, a study of state changes can enrich your understanding of the system. For example, you can look at any state and ask whether you may have missed a possible state change. That is the case in Table 6.21 because it is certainly possible to go from the state "loaded, locked, armed" to the state "unloaded, locked, armed." That would correspond to the child removing the projectile from the catapult when it is in the locked and armed state. We would call that event "unload" and place it in the appropriate cell. We missed capturing that state change in Table 6.21 because it did not show up in any of the operational descriptions. It seems a rather trivial state change, so we have decided to ignore it.

As with the functional interrelationship matrix, the state change matrix is an example of a precedence matrix. The cell entries indicate that the row state precedes the column state at the time of a particular event. In Table 6.21, it is impossible to find an ordering of rows and columns that puts all of the entries above the diagonal. That indicates that there are loops in the sequences. (The converse is also true: If you can find a loop, then you know that it will be impossible to get all entries above the diagonal.) Table 6.22 identifies one such infinite loop: from state "unloaded, locked, armed" through event "misfire" to state "unloaded, unstable, unarmed" through event "receptacle in position" to state "unloaded, locked, armed" through event "receptacle secured" back to state "unloaded, locked, and armed." For each such loop, you should ask the question if it represents an undesirable behavior, such as feedback gain or program crash. In each case for the toy catapult, however, the user (the child) must make a positive action to continue the behavior, so the loop is completely within the user's control. It is not undesirable.

Table 6.21   Summary of State Changes

(Event) Changes State From, to	Unloaded, Stable, Unarmed	Unloaded, Unstable, Unarmed	Unloaded, Locked, Unarmed	Unloaded, Locked, and Armed	Loaded, Locked, and Armed	Loaded, Released (Unstable), and Unarmed	Loaded, Stable, Unarmed
Unloaded, stable, unarmed		"energy in"					
Unloaded, unstable, unarmed	"abort"		"receptacle in position"				
Unloaded, locked, unarmed				"receptacle secured"			
Unloaded, locked, armed		"misfire"			"load"		
Loaded, locked, armed						"release action"	
Loaded, released (unstable), unarmed							"stop"
Loaded, stable, unarmed	"launch"						

Table 6.22 An Infinite Loop of State Changes

(Event) Changes State from, to	Unloaded, Stable, Unarmed	Unloaded, Unstable, Unarmed	Unloaded, Locked, Unarmed	Unloaded, Locked, Armed	Loaded, Locked, Armed	Loaded, Released (Unstable), Unarmed	Loaded, Stable, Unarmed
Unloaded, stable, unarmed		"energy in"					
Unloaded, unstable, unarmed	"abort"		"receptacle in position"				
Unloaded, locked, unarmed				"receptacle secured"			
Unloaded, locked, armed		"misfire"			"load"		
Loaded, locked, armed						"release action"	
Loaded, released (unstable), unarmed							"stop"
Loaded, stable, unarmed	"launch"						

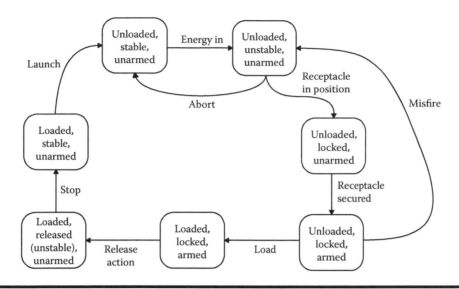

Figure 6.4   State change diagram.

There is a diagrammatic view of state changes that is equivalent to the state change matrix. The *state change diagram* displays states as rounded rectangles and connects them with curvilinear arcs representing the events that mark the transition from one state to another. Figure 6.4 is a state change diagram that is equivalent to the state change matrix of Table 6.21. From this diagram, it is easy to spot three simple loops, or cycles. A *simple cycle* is a path through the network that connects a node with itself, but does not visit any other node more than once along the path. One simple cycle involves the "abort" event, another involves the "misfire" event, and a third involves the "load" event. If we were to include the "unload" event described earlier, then this would reveal a fourth simple cycle. It is easy to check that each of these simple cycles involves some sort of user action to continue. Thus, all of the cycles can be seen to be under user control.

So far, we have described two abstract ways of viewing the behavior of systems: the functional view and the state change view. These views have the advantage of compressing the essence of behavior from many pages of detail to matrices, or diagrams, that fit on a single page. Many people are comfortable creating designs at this abstract level without resorting to the detailed behavioral view. We have emphasized the behavioral detail first because it naturally integrates the two abstract views. You can always get to this level of abstraction by starting with the detailed behavior (diving) and then summarizing the detail (surfacing) using the matrix approach of this chapter.

Design the Structure

In this section we focus on the physical architecture of the system: the main subsystems and their interactions. We do this in six steps:

1. Identify interfaces and finalize subsystems.
2. Document links between subsystems.
3. Identify emergent interactions.

4. Develop a rough-cut bill of materials.
5. Estimate material cost and system reliability.
6. Allocate target technical performance measures to subsystems.

Identify Interfaces and Finalize Subsystems

We summarize the system interfaces by means of an interface matrix. An *interface matrix* is a square matrix formed by listing the system entities, both internal and external, as the row and column headings and by using the cells to describe the nature of the interface between each pair of entities. In our case, we use it to describe the signal or message transmitted from one subsystem to another. To illustrate, Table 6.23 lists all of the internal subsystems and some of the external entities for the toy catapult and names the event or message that enables one subsystem to trigger a function (or change the state) in another subsystem. Every interface that is identified typically requires some form of documentation (often called an *interface control document*). More importantly, someone on the design team is assigned responsibility for managing the design interface.

Observe that the interface matrix is a summarization of all the interface rows that were identified in the operational descriptions.

A *design structure matrix* is an interface matrix using only internal entities for the rows and columns. Table 6.24 is the design structure matrix that results from deleting the rows and columns of external entities from Table 6.23. If you imagine that different design teams are responsible for different subsystems, then the design structure matrix identifies interfaces where design decisions of one team can have a critical impact on the design success of another team. It frequently happens that teams make different assumptions about the nature of the interface. The result is often a failure of the system, usually discovered when the subsystems are first integrated. It is a maxim that "systems tend to fail at the interfaces" for this reason.

One of the techniques of dealing with design structure matrices is to reorder the rows and columns to *cluster the subsystems* into groups that exhibit high degrees of interface within the group but low degrees of interface with other groups. Mathematically, we seek to form a block diagonal matrix where most of the entries in the matrix exist within blocks along the diagonal and only a few entries exist outside these blocks. There are algorithms from computer science capable of automatically suggesting how best to reorder such matrices. In the spreadsheets that accompany this text, we provide a collection of such algorithms coded in Visual Basic macros. It is a simple matter to copy a design structure matrix into one of these spreadsheets and then reorder it using any of these macros.

Table 6.25 shows the results of a reordering macro applied to the design structure matrix of Table 6.24. The suggested blocks along the diagonal are shown in heavy outline. The reordering reveals that the locking system is tightly connected to the trigger system. Because of the degree of interconnection, it is advisable for the design teams responsible for these subsystems to work closely together or, if possible, for both subsystems to be assigned to a single design team. Accordingly, from this point onward we no longer refer to these as separate subsystems; rather, we rename them "the lock and trigger system."

Document Links

The interfaces have been identified and the subsystems finalized, it is now important to document the different methods of transmission across the interfaces. In particular, we use the term *link* to

Table 6.23 Interface Matrix for Toy Catapult

(Event) Relates Subsystem to	Operator (Child)	Toy					Projectile	Pet Rodent
		Energy Storage System	Locking System	Projectile Containment System	Trigger System	Projectile Launch System		
Operator (child)		"energy in"		"load"	"trigger"		"load"	"load"
Energy storage system	"maximum energy reached"					"energy out"		
Locking system		"receptacle secured," "release action"			"receptacle secured"			
Projectile containment system								
Trigger system			"release command"					
Projectile launch system				"stop"			"launch"	"launch"
Projectile								
Pet rodent				"escapes"		"survives"		

Table 6.24   Design Structure Matrix for the Toy Catapult

		Toy				
	(Event) Relates Subsystem to	Energy Storage System	Locking System	Projectile containment system	Trigger System	Projectile Launch System
Toy	**Energy storage system**					"energy out"
	Locking system	"receptacle secured," "release action"			"receptacle secured"	
	Projectile containment system					
	Trigger system		"release command"			
	Projectile launch system			"stop"		

refer to the physical means of implementing the interface. In electromechanical systems, these links may be serial, parallel, Ethernet, or wireless connections, and the details must be established early so that it is not left to assumption by the different design teams. In the case of the toy robot, none of the links will be electrical. They will all be mechanical devices of some sort.

Table 6.26 summarizes what is known about the links in the toy catapult design. You can see that we are too vague in our descriptions at this stage. It would be impossible for different design teams to work separately on these subsystems until the links were more explicitly designed. For example, if a team wanted to design the receptacle (the projectile containment system), it would need to know the dimensions of the catapult arm. The link matrix, Table 6.26, becomes a critical tool in interface design.

Identify Emergent Interactions

By this point in the text, it should be clear that the matrix approach is the single most powerful technique we have offered to thinking systematically about relationships that are important to design. To this point, we have considered functional interrelationships, state changes, subsystem interfaces, and links. These relationships have flowed naturally from a consideration of use case behaviors and functional analysis. However, other relationships must be captured, and these can be subtler to discover. We refer here to *emergent interactions*—that is, unintended side effects that the operation of one subsystem may have on other subsystems.

Table 6.25   Reorder and Regroup Subsystems

(Event) Relates Subsystem to	Projectile Launch System	Energy Storage System	Projectile Containment System	Lock and Trigger System	
				Locking System	*Trigger System*
Projectile launch system			"stop"		
Energy storage system	"energy out"				
Projectile containment system					
Locking system		"receptacle secured," "release action"			"receptacle secured"
Trigger system				"release command"	

(left margin label: Lock and trigger system)

An example of an emergent interaction in the toy catapult would be the influence that the projectile launch system may have on the reliability of other subsystems. For example, it is likely that the launch system will require a hard stop in order to cause the projectile to continue its trajectory and leave the receptacle. This hard stop will likely cause vibrations. These vibrations will likely be transmitted through the toy structure to other subsystems, such as the lock and trigger system. The design team responsible for the lock and trigger system may design a delicate mechanism that works well in the laboratory, but fails the rigors of testing when integrated with the rest of the toy. The team may not have anticipated the effects of vibration.

Emergent interactions are common in electromechanical systems; they are less common in software systems. Heat, friction, vibration, static electricity, magnetic fields, radio interference, corrosion, acidity, alkalinity, grease, dirt, and contamination are all examples of side effects from some operations that can be harmful to other subsystems. A systematic way to discover and communicate such interactions is with an *interaction matrix,* which is simply a square matrix of the system entities (internal and external) in which the cells identify the potential for an emergent interaction. Table 6.27 displays an interaction matrix for the toy catapult in which we noted

Table 6.26   Define Links

(Event) Relates Subsystem to	Operator (Child)	Projectile Launch System	Energy Storage System	Projectile Containment System	Lock and Trigger System
Operator (child)			manual contact (push down)	manual contact (load)	
Projectile launch system				mechanical connection (catapult)	mechanical
Energy storage system		mechanical energy transfer (pivot)			mechanical
Projectile containment system					
Lock and trigger system			mechanical release		

the potential problem of vibration. This was also an opportunity to note that the weight of the projectile containment system (the receptacle) has a significant design influence on the energy storage system. If the weighted catapult concept is used, the design of the mass of the weight will have to take the mass of the receptacle into account.

What should be done once an emergent interaction is discovered? This will typically give rise to requirements that will be imposed on individual subsystems. For example, we could impose a requirement on the projectile launch system to isolate the effect of vibration from the rest of the catapult. This could be achieved, for example, by using a soft stop (a padded crossbar) instead of a hard stop. Alternatively, we could require the other subsystems, such as the lock and trigger system, to be designed to satisfy their reliability targets in the presence of certain levels of vibration. Thinking creatively, we could require the projectile containment system to eject its projectile without a hard stop (perhaps using a slingshot concept fragment). In some cases, whole subsystems must be introduced just to deal with emergent interactions. Cooling subsystems, for example, are critically important in many applications to cope with the side effects of friction. Whatever the decision, the consideration of emergent interactions is a major source of new derived requirements.

The existence of an interaction may also require different design teams to work more closely together. Considering the connection between the projectile containment system and the energy storage system that is revealed in Table 6.27, we may want to revisit Table 6.25. As suggested by the block diagonals in Table 6.25, it may be wise to limit the design organization to two teams: one team to focus on the lock and trigger system and the other team to focus on the other subsystems.

Table 6.27 Identify Emergent Interactions

(Row) Negatively Affects (Column) by	Operator (Child)	Projectile Launch System	Energy Storage System	Projectile Containment System	Lock and Trigger System	Projectile	Pet Rodent
Operator (child)							
Projectile launch system				vibration	vibration		
Energy storage system							
Projectile containment system			weight acts as counter-balance				
Lock and trigger system							
Projectile							
Pet rodent							

Calling the Air Force to Task

A high level task force in 2008 criticized the Air Force for failing to follow basic systems engineering processes. The task force identified a number of programs with costly failures.

Among them [was a] military satellite system designed to detect foreign missile launchings that [the leader of the task force] said was inexplicably designed with two sensors that cannot operate simultaneously on the same spacecraft without extensive, costly shielding to prevent electromagnetic interference generated by one from disabling the other.

Philip Taubman, *New York Times,* **June 25, 2008**

Sketch a Design Concept

Before proceeding into detailed subsystem design, we need a more complete integrated design concept. Concept sketches that show the proposed structure of the system, even at a crude or preliminary level, are useful at this stage. Based on our exploration and optimization cycles for the toy catapult design, we selected the weighted catapult concept for the basic system concept (Figure 6.5), the pivot ring concept for the lock and trigger system (Figure 6.6), and a ball-throwing pitcher concept for the enclosure design (Figure 6.7 and Figure 6.8). The idea of the enclosure is to protect the child from the moving parts of the catapult as much as possible. The sides of the

Figure 6.5   Basic system concept.

enclosure can be used to give the toy a personality, such as "Peter the Pitcher." The image for the pitcher is inspired by the famous profile of Baseball Hall of Fame pitcher Juan Marichal of the San Francisco Giants. The brainstorming and concept selection steps that led to the enclosure design have been omitted from the text.

Create a Rough-Cut Bill of Materials

Based on the product concept sketch, we can create a rough-cut bill of materials. A *bill of materials* is a listing of the components, with quantities and estimated cost, that must be purchased or manufactured in order to assemble the product. The bill is typically ordered into a hierarchy of system and subsystems. Table 6.28 displays an indented bill of materials for the toy catapult. Each new indentation level suggests a new level of subsystem. Notice how the numbering system follows the indentation levels. Part 1.4.2 is at the third level of indentation; part 1.4.2.1 is at the fourth level.

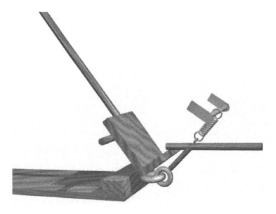

Figure 6.6   Lock and trigger concept.

Figure 6.7   Enclosure concept: front view.

This may not be the final design, but it will prove useful in the next section to have dived into this level of detail because we can estimate material costs at the component level in a relatively straightforward process. We step through each component, estimate the type of raw material that will be required, establish units of measure (metric units, in our case), estimate the quantity of material required per toy, and estimate the material cost per unit. From there, it is a simple matter

Figure 6.8   Enclosure concept: side view.

Table 6.28   Indented Bill of Materials with Hierarchical Numbering

1	*Toy Catapult*			
1.1		Projectile launch system		
1.1.1			Arm	
1.1.2			Pivot	
1.2		Energy storage system		
1.2.1			Pendulum weight	
1.2.2			Bracket	
1.2.3			Base	
1.3		Projectile containment system		
1.3.1			Receptacle	
1.4		Lock and trigger system		
1.4.1			Locking system	
1.4.1.1				Latch
1.4.1.2				Latch support
1.4.1.3				Extension spring
1.4.1.4				Eye rings (2)
1.4.1.5				Spring bar
1.4.2			Trigger system	
1.4.2.1				Trigger arm
1.5		Body		
1.5.1			Front cover	
1.5.2			Back cover	
1.5.3			Braces (2)	
1.5.4			Fasteners	

to compute the component material cost per toy (= estimated quantity per toy × material cost per unit), and sum the component costs to get subsystem and system totals. Table 6.29 illustrates these calculations for the toy catapult.

Material cost is a major component of the final cost of a product, often exceeding 70% of the manufactured cost. It is important to get an estimate of the material cost early in the product development phase. Many techniques can be used for cost estimation. The text *Engineering Cost Estimating* (Ostwald 1991) is an excellent reference.

Create a Rough-Cut Reliability Estimate (Advanced Topic)

The detailed bill of materials also permits us to estimate the reliability of the system. Reliability is typically expressed in terms of *mean time to failure* (MTTF): the average number of operating hours before a failure takes place. Note that the expression involves operating hours rather than real time. Failures tend to be correlated with usage and not just age. If the toy catapult is not played with for months at a time, it will last longer than if it is played with daily. MTTF refers to the expected life of the toy in terms of hours used. For calculation purposes, it is easier to work with failure rates rather than MTTF. The *failure rate* of a system or component is the number of failures expected per unit of operating time. The failure rate is the inverse of the MTTF:

$$Failure\ rate = \frac{1}{\text{MTTF}}.$$

Estimating the MTTF of a system is a little more complicated than cost estimation because we must convert back and forth between MTTF and failure rates. For the purpose of a rough-cut reliability estimate, we assume that all failures are independent of each other and that the failure rate of a subsystem is simply the sum of the failure rates of its components (we assume a failure in any component will cause the system to fail). Therefore, the estimation process is to estimate first the MTTF of the components, convert them to failure rates, sum the failure rates by subsystem and by system, and then convert back to MTTF.

Table 6.30 illustrates for the toy catapult. Observe that we began by estimating the MTTF, in hours of play, for each component of the rough-cut bill of materials. In general, component manufacturers will provide estimates of MTTF for their products. When such estimates are not available, some testing laboratories can estimate component lifetimes by testing batches of components (usually by running all the components of the batch for a specified length of time and counting the number of failures). For a hobbyist project such as the toy catapult, we are forced to make our own estimations. We felt that many of the structural components would not fail from normal use. We used the MTTF of 10,000 hours to represent these items. The components of the lock and trigger system are more likely to experience failure, particularly if they are small wooden pieces. We estimated shorter lives for each of these components, with the extension spring and trigger arm potentially being the weakest components.

Once the MTTFs of the components are all estimated, it is simple arithmetic from that point onward. In the next column, we invert the MTTF to get failure rates. We then compute subtotals of failure rates by subsystem and total the failure rate of the system (0.07623 failures per hour). From these subtotals and totals, we invert back to get the MTTF of each of the subsystems and of the system as a whole (13 hours). Observe that although the average MTTF across components

Table 6.29 Material Cost Estimation for Toy Catapult

Bill of Materials		Raw Material Description	Unit of Measure	Estimated Quantity per Toy	Material Cost per Unit	Material Cost per Toy	Subtotals	Subsystem Totals	System Total
1	Toy catapult								$25.05
1.1	Projectile launch system						$0.73	$0.73	
1.1.1	Arm	Wood 1 in. × 0.75 in.	Centimeter	20	$0.030	$0.600			
1.1.2	Pivot	Steel rod	Centimeter	5	$0.025	$0.125			
1.2	Energy storage system						$8.60	$8.60	
1.2.1	Pendulum weight	Steel	Gram	50	$0.002	$0.100			
1.2.2	Bracket	Steel U-shape with predrilled holes	Count	1	$1.000	$1.000			
1.2.3	Base	Wood 0.5 in. thick	Square centimeter	300	$0.025	$7.500			
1.3	Projectile containment system						$1.50	$1.50	
1.3.1	Receptacle	Wood 0.25 in. thick	Square centimeter	75	$0.020	$1.500			
1.4	Lock and trigger system							$1.02	
1.4.1	Locking system						$0.52		
1.4.1.1	Latch	Wood dowel 0.25 in diameter	Centimeter	5	$0.020	$0.100			

1.4.1.2	Latch support	Wood dowel 0.25 in diameter	Centimeter	5	$0.020	$0.100		
1.4.1.3	Extension spring	Steel spring 0.25 lb force	Count	1	$0.100	$0.100		
1.4.1.4	Eye rings (2)	Steel eye ring	Count	2	$0.050	$0.100		
1.4.1.5	Spring bar	Steel pin	Centimeter	6	$0.020	$0.120		
1.4.2	Trigger system						$0.50	
1.4.2.1	Trigger arm	Wood dowel 0.25 in diameter	Centimeter	25	$0.020	$0.500		
1.5	Body						$13.20	$13.20
1.5.1	Front cover	Wood 0.25 in. thick	Square centimeter	300	$0.020	$6.000		
1.5.2	Back cover	Wood 0.25 in. thick	Square centimeter	300	$0.020	$6.000		
1.5.3	Braces (2)	Wood dowel 0.25 in diameter	Centimeter	10	$0.020	$0.200		
1.5.4	Fasteners	Assorted screws	Count	20	$0.050	$1.000		

Table 6.30 Rough-Cut Estimate of Toy Catapult Reliability

Bill of Materials		Estimated Mean Time to Failure (MTTF) (Hours of Play)	Failure Rates (Failures per Hour)	Failure Rate Subtotals (Failures per Hour)	Failure Rate Subsystem Totals (Failures per Hour)	Failure Rate System Total (Failures per Hour)	MTTF Subsystems (Hours of Play)	MTTF System (Hours of Play)
1	Toy catapult					0.07623		13
1.1	Projectile launch system			0.00110	0.00110		909	
1.1.1	Arm	10,000	0.0001					
1.1.2	Pivot	1,000	0.001					
1.2	Energy storage system			0.00030	0.00030		3,333	
1.2.1	Pendulum weight	10,000	0.0001					
1.2.2	Bracket	10,000	0.0001					
1.2.3	Base	10,000	0.0001					
1.3	Projectile containment system			0.00010	0.00010		10,000	
1.3.1	Receptacle	10,000	0.0001					
1.4	Lock and trigger system				0.07433		13	
1.4.1	Locking system			0.05433				
1.4.1.1	Latch	75	0.013333333					

ID	Name					
1.4.1.2	Latch support	100	0.01			
1.4.1.3	Extension spring	50	0.02			
1.4.1.4	Eye rings	1,000	0.001			
1.4.1.5	Spring bar	100	0.01			
1.4.2	Trigger system			0.02000		
1.4.2.1	Trigger arm	50	0.02			
1.5	Body			0.00040	0.00040	2,500
1.5.1	Front cover	10,000	0.0001			
1.5.2	Back cover	10,000	0.0001			
1.5.3	Braces	10,000	0.0001			
1.5.4	Fasteners	10,000	0.0001			

Table 6.31 Target TPMs for Reliability and Material Cost

OR.17	The mean cycles to failure of the toy catapult shall be at least 1,000.	Mean cycles to failure
OR.22	The material cost of the toy catapult shall not exceed $20.	Material cost

is in the thousands of hours, the MTTF of the system is only 13 hours, which is even lower than MTTF of the least reliable component. This illustrates two ideas in the study of reliability: (1) The system is only as good as the weakest components (the extension spring and trigger arm in this case), and (2) the more components a system requires, the more likely it is that some required component will fail, causing the system to fail.

Allocate Target Technical Performance Measures to Subsystems

Recall that the originating requirements from the house of quality analysis included specific targets for reliability and material cost (Table 6.31). We need to allocate these target technical performance measures (TPMs) to the individual subsystems so that each design team has a target in each of these categories. The goal is for each subsystem to satisfy targets so that when the system is integrated, the overall target TPMs are met. In general, this allocation process must be performed for every TPM. We illustrate the allocation process for material cost and reliability only.

Allocation of Target Cost to Subsystems

It will frequently be the case that the target TPMs are more severe than the rough-cut estimates. For example, we estimated the material cost of the toy catapult to be $25.05, but requirement OR.22 calls for a target material cost of $20. We must achieve a 25% reduction in cost to meet this target. One simple way of allocating the cost target to the subsystems is to compute the fraction of rough-cut cost accounted for by each subsystem and then multiply that fraction by the target cost to get the target cost of each subsystem. Table 6.32 shows the calculations. Observe that because the body accounted for 52.7% of the rough-cut cost ($13.20 as a fraction of $25.05), it receives 52.7% of the target cost. We look, therefore, to the team responsible for designing the body to make the greatest reduction in material costs.

You can imagine situations in which this allocation mechanism is too simplistic. For example, because of the importance of the lock and trigger system from a reliability perspective, we may want to increase the target cost for that subsystem. This would permit that design team to use more reliable materials. To achieve the overall system cost, we would then need to find other, less critical subsystems, such as the body, in which to make even greater cost reductions. In general, the allocation of TPMs to subsystems is a process of negotiation between the system architect (the designer responsible for the overall system) and the subsystem design team leaders. If there are known degrees of design freedom, then it might be formulated as an optimization problem and solved mathematically.

Table 6.32   Allocation of Target Cost to Subsystems Based on Rough-Cut Cost Distribution

		Subsystem Totals	*System Total*	*Percent Distribution*	*System Target Cost*	*Subsystem Allocated Target Costs*
1	Toy catapult		$25.05		$20.00	
1.1	Projectile launch system	$0.73		2.9		$0.58
1.2	Energy storage system	$8.60		34.3		$6.87
1.3	Projectile containment system	$1.50		6.0		$1.20
1.4	Lock and trigger system	$1.02		4.1		$0.81
1.5	Body	$13.20		52.7		$10.54

Allocation of Target Reliability to Subsystems (Advanced Topic)

In a similar fashion, we allocate the reliability TPM to the different subsystems. We must first reconcile the units of measure. Requirement OR.17 calls for a target mean cycles to failure of 1,000 cycles. We expressed the rough-cut reliability estimate as 13 hours. Suppose we estimate the number of play cycles per play hour at 20. That allows 3 minutes for each cycle (load, launch, and fetch projectile). With that assumption, requirement OR.17 can be re-expressed as "the system shall achieve an MTTF of 50 play hours." As in the case of the cost target, the target reliability TPM is more severe than the rough-cut estimate—in this case by a factor of four. We are going to have to make significant improvements to at least one of the subsystems to achieve this new target.

Suppose we proceeded blindly, as in the case of the cost allocation, and allocated the target reliability (expressed as failure rate) based on the fraction of failure rate that each subsystem accounted for. Table 6.33 shows the calculations. We first convert the rough-cut MTTF numbers back to failure rates. Then, because the failure rate for the lock and trigger system is 97.5% of the total system failure rate, we apply that fraction to the target system failure rate of 0.02 failures per hour (the inverse of the target MTTF of 50) to get an allocated failure rate of 0.019502; when inverted, this yields a target MTTF of 51. The calculations for the other subsystems follow the same pattern. Therefore, this calls for a dramatic improvement in the reliability of the lock and trigger system from the rough-cut estimate.

Table 6.33 reveals a problem with the simplistic approach. Recall that we intended 10,000 hours to represent a highly reliable subsystem. As a result of the simple allocation approach, we now have a target MTTF of almost 40,000 hours for the projectile containment system. This

Table 6.33 Simplistic Target Reliability Allocation

		MTTF Subsystem (Hours)	MTTF System (Hours)	Failure Rate Subsystem Totals (Failures/Hour)	Failure Rate System Total (Failures/Hour)	Percent Distribution	System Target MTTF (Hours)	System Target Failure Rate (Failures/Hour)	Subsystem Allocated Failure Rates (Failures/Hour)	Subsystem Allocated MTTF (Hours)
1	Toy catapult		13		0.07623		50	0.02		
1.1	Projectile launch system	909		0.0011		1.4			0.000289	3,465
1.2	Energy storage system	3,333		0.0003		0.4			0.000079	12,706
1.3	Projectile containment system	10,000		0.0001		0.1			0.000026	38,117
1.4	Lock and trigger system	13		0.0743		97.5			0.019502	51
1.5	Body	2,500		0.0004		0.5			0.000105	9,529

is excessive for a hobbyist project. Another approach might be to focus on the lock and trigger system, asking the question, "If the failure rates for the other subsystems are unchanged, what must the failure rate of the lock and trigger system be in order to achieve the target failure rate?" Table 6.34 reveals the answer. After subtracting the failure rates of the other subsystems from the target failure rate, we see that the lock and trigger system cannot exceed a failure rate of 0.0181 failures per play hour. Converting back to MTTF, we see that the lock and trigger system must achieve a target MTTF of 55 hours. The targets in Table 6.34 are more reasonable than the ones in Table 6.33.

No single approach to allocating TPMs is satisfactory in all situations. You will need to devise your own rules, based on the specifics of the situation.

What can be done to improve the reliability of a subsystem? There are two basic strategies. One is to increase the reliability of the components (stronger materials, more robust designs), and the other is to include redundant components (one component continues to provide the function even if another fails). When systems include redundancy, the calculations for system reliability are more complicated. Such calculations are beyond the scope of this text. The approach of this chapter is sufficient for nonredundant systems.

Dive and Surface in TPM Allocation

In this approach to allocating target TPMs, you can see again the diving and surfacing analogy. Although our basic goal is abstract (allocating TPMs to subsystems), it is easier to approach that task by first diving into detail (estimating rough-cut bill of materials cost and system reliability) before surfacing and addressing the more abstract issue. That way, our allocation is not arbitrary because it has some basis in a detailed view of the system. We have not fixed the design for any of the subsystems. The team responsible for the lock and trigger system may continue to use the pivot ring concept, but it is also free to explore other design ideas. What we have fixed is the target material cost and the target reliability for each of the subsystems. For example, the lock and trigger system team is free to redesign the subsystem, provided that it meets the material cost target of $0.81 and the MTTF target of 55 hours.

We have shown the allocation of target TPMs to subsystems for cost and reliability. In general, many such allocation problems must be solved: weight restrictions, power limitations, communication bandwidth, and so on. In each case, we suggest using a dive-and-surface approach: a bottom-up, rough-cut estimate of need followed by a top-down allocation of budget guided by the estimates of need.

Summary

In this chapter we did not advance the design of the toy catapult by very much. That is not the purpose of developing an architecture. What we did accomplish was to break up the overall design problem into a problem of designing subsystems. We described the system architecture—a critical skill to acquire if you are going to use team approaches to design. Using operational description templates, we described the behavior of each use case in detail, showing how the different subsystems must cooperate to accomplish the high-level behavior.

The use of interface rows and state change columns in these templates permits a very granular view of behavior. The operational description templates then become the basis for writing operator

Table 6.34 Reliability Allocation Gain Focused on Lock and Trigger System

		MTTF Subsystem (Hours)	MTTF System (Hours)	Failure Rate Subsystem Totals (Failures/Hour)	Failure Rate System Total (Failures/Hour)	System Target MTTF (Hours)	System Target Failure Rate (Failures/Hour)	Subsystem Allocated Failure Rates (Failures/Hour)	Subsystem Allocated MTTF (Hours)
1	Toy catapult		13		0.07623	50	0.02		
1.1	Projectile launch system	909		0.0011				0.001100	909
1.2	Energy storage system	3,333		0.0003				0.000300	3,333
1.3	Projectile containment system	10,000		0.0001				0.000100	10,000
1.4	Lock and trigger system	13		0.07433				0.018100	55
1.5	Body	2,500		0.0004				0.000400	2,500

interface specifications, hardware system functional specifications, software system functional specifications, test procedure specifications, and interface specifications. In particular, the functional specifications for individual subsystems can be read from the templates by simply "walking the column" for that subsystem and collecting the functional requirements. The detail captured by the operational description templates can be summarized in different ways: a functional interrelationship matrix, a state change matrix, and an interface matrix.

Each of these summary matrix views also has an equivalent diagrammatic view. We used matrix operations of reordering rows and columns to improve the presentation of these matrices. In particular, the clustering of the interface matrix led us to assigning two subsystems (the locking system and the trigger system) to a single design team because of the high degree of interconnectedness of these subsystems. We used a matrix approach to identify physical links between subsystems that needed to be defined more completely. We also used a matrix approach to identify emergent interactions between subsystems that could suggest additional design requirements and team assignments. Finally, we allocated system-level target technical performance measures to subsystems in a two-step approach: a bottom-up estimation of current design performance followed by a top-down allocation of targets.

Discussions

1. What process would have to be in place to avoid mistakes such as the Mars Climate Orbiter mishap?
2. Prior to the invention of the Post-it note, one of the proposed uses for Silver's glue was to cover a bulletin board with it so that papers could be posted without tacks or pins. Discuss the link between this idea and the Post-it note. Why was the connection not made immediately?
3. Pick any common device, such as an automobile, and identify major components, such as cooling systems, that exist to shield the operator, the environment, or other components from negative side effects of the primary subsystems (the engine and power train in the case of an automobile). Estimate the fraction of the total cost of the device that can be traced to these side effects.

Exercises

1. Create operational description templates for three use cases involving at least four subsystems (internal entities) that must work together to achieve the behavior. For example, the BathBot system will have subsystems such as a sensor system, navigation control subsystem, flexible arm, end effectors (brush, wiping pads, spray nozzle), and a central controller. Use the following checklist to make sure your template is complete:
   a. behavior name;
   b. initial conditions;
   c. column for operator;
   d. columns for internal subsystems;
   e. columns for external entities;
   f. rows for functional requirements (stated in the language "The system shall...");

g. interface rows (with event or message names) whenever subsystems must cooperate; choose a different background color for these cells;

h. ending conditions;

i. notes;

j. system state column; and

k. timing target.

2. Use the operational description templates you created in the previous exercise and summarize them with the following:

a. a functional interrelationship matrix; and

b. a state change matrix.

3. Extend your analysis of the system you studied in the previous exercise by creating an interaction matrix. You are looking for potentially negative ways in which the subsystems could interact with each other.

4. Recommend a target MTTF for each of the toy catapult's subsystems so that the system as a whole achieves a target MTTF of 60 hours and no subsystem is required to have an MTTF in excess of 10,000 hours (download and use the spreadsheet "Toy08 Bill of Materials.xls").

References

Bass, L., P. Clements, and R. Kazman. 2003. *Software architecture in practice*, 2nd ed. Reading, MA: Addison–Wesley.

Karas, L. 2001. The systems engineering environment (SEE) tutorial. International Council on Systems Engineering Conference.

———. 2006. Owego systems engineering principles and practices course. Lockheed Martin Systems Integration—Owego.

Karas, L., and D. Rhodes. 1987. Systems engineering technique. NATO Advisory Group for Aerospace Research & Development Conference Proceedings no. 417.

———. 1993. Enabling concurrent engineering through behavioral analysis. Proceedings of the Third International Council on Systems Engineering, July.

———. 1999. The common denominator. Proceedings of the Ninth International Council on Systems Engineering, June.

Maier, M. W., and E. Rechtin. 1997. *The art of systems architecting*. New York: CRC Press.

MIT School of Engineering. Inventor of the week archive: Art Fry and Spencer Silver. http://web.mit.edu/invent/iow/frysilver.html (accessed December 16, 2008).

NASA. 1999. Mars Climate Orbiter Mishap Investigation Board Phase I Report.

NASA/JPL-Caltech. Mars Climate Orbiter. Artist's conception. http://www.jpl.nasa.gov/missions/mission-Images.cfm?missionID=179 (accessed January 2, 2009).

Ostwald, P. 1991. *Engineering cost estimating*, 3rd ed. New York: Prentice Hall.

Spinrad, R.J. As quoted in Maier and Rechtin, 1997.

Taubman, P. 2008. Top engineers shun military; concern grows. *New York Times*, June 25.

von Braun, W. As quoted in Maier and Rechtin, 1997.

Chapter 7

Validate the Design

If you think your design is perfect, it's only because you haven't shown it to someone else.

Harry Hillaker, F-16 architect

Every once in a while you have to go back and see what the real world is telling you.

Harry Hillaker

The phrase, "I hate it," is direction.

Lori I. Gradous

Introduction

Verification and validation (V&V) is the process of ensuring that the product or system being built satisfies the requirements for which it is being designed and that it will satisfy the customer for whom it is being built. In general, the word "validate" is intended to mean "build the right product"—that is, "build the product the customer needs." The word "verify" is intended to mean "build the product right"—that is "conduct each step of the design and build process correctly so that the integrated product works correctly." Another way of looking at it is that "validation" looks at how your customer will evaluate your product or service, whereas "verification" refers to how you will test your product or service. In practice, these meanings are often blurred. We take the word "validate" to include the process of bringing the design project to a successful conclusion, whereas the word "verify" has a narrower meaning of checking each step.

Figure 7.1 is a "Vee" diagram that shows how design should always be connected with test. Our focus is not on the engineering design–build–test sequence below the line. These activities are often subcontracted. Instead, in Getting Design Right, we focus on the design and test that occur above the line—that is, before and after this "inner V." Ours is more of a systems view of design.

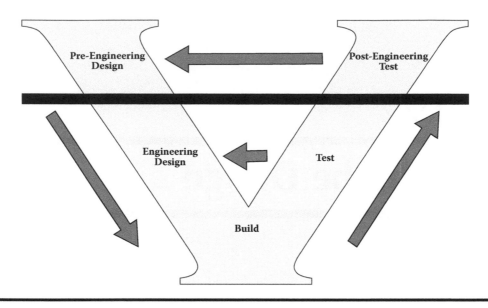

Figure 7.1   The "Vee" diagram: linking test to design.

If you are responsible for a design project, particularly a project that involves a team or teams of people, then you will want to know what steps you can take as a *design project manager* to increase the likelihood of project success. This chapter focuses on two major areas of your responsibility that contribute strongly to project success. Expressed as steps in validating the design, they are to

1. verify requirements; and
2. manage design risks.

The Hubble Space Telescope Test Failure

If you happen to visit the "Explore the Universe" exhibit at the Smithsonian Institution's National Air and Space Museum in Washington, D.C., you can see the Hubble Space Telescope's never-used backup mirror on display. The mirror was fabricated by the Eastman Kodak Company in Rochester, New York, and completed in 1980 several months before the telescope's primary mirror was unveiled. The Kodak mirror was correctly ground but never coated.

In the late 1970s, both the Perkin–Elmer Corporation and Eastman Kodak bid for the opportunity to manufacture this principal feature of the Hubble Space Telescope. The Perkin–Elmer bid was two-thirds the price of the Eastman Kodak mirror, but it had one critical difference: There would be no final assembly test before launching the telescope into space. Such a test would have cost an additional $100 million, and both NASA and Perkin–Elmer were convinced that the final assembly test would not be needed (AP, 1990).

The Eastman Kodak proposal, however, called for the production of two identical primary mirrors: one in Rochester and the other at the Itek Optical System plant in Lexington, Massachusetts. The mirrors would then undergo comparative testing,

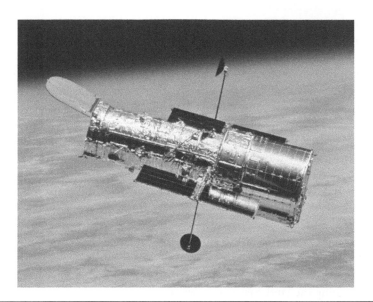

Figure 7.2   The Hubble Space Telescope. (Source: NASA.)

which would ensure that the Hubble telescope did not leave the ground with any unnoticed optical flaws. "We have never had a satellite system with an optical system remotely similar to Hubble without a final assembly test," remarked then Senator Al Gore, 1992 chairman of the Senate Commerce Subcommittee on Science, Technology, and Space (Leary, 1990). Nevertheless, Perkin–Elmer won the bid to construct the primary mirror, and Eastman Kodak was subcontracted to provide the backup.

Several weeks after the $1.5 billion telescope was sent into space (Figure 7.2) in 1990, it was discovered that the mirror contained a severe focusing defect, so the

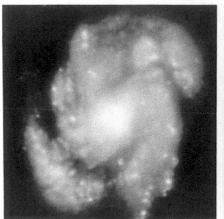
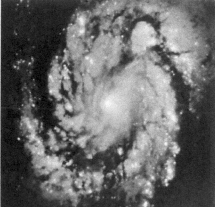

Wide Field Planetary Camera 1   Wide Field Planetary Camera 2

Figure 7.3   Improvement in Hubble images after corrective optics were installed, 1993. (Source: NASA.)

images returning to earth from Hubble were blurred (Figure 7.3). Due to a miscalibrated measuring instrument at the Perkin–Elmer plant, the primary mirror had been ground and polished to the wrong shape. Though the error was less than 3/1000 of a millimeter, it was enough to render the telescope unfit for the deep space observations in its mission (U.S. History Encyclopedia). By this point, it was impossible to switch the faulty primary mirror with its accurate backup (Figure 7.4).

"It's wild," said Dr. Frank D. Drake, an astronomer at the University of California at Santa Cruz. "There's a perfect mirror sitting in a crate somewhere. Any ground-based observatory that had two mirrors in preparation would have compared the two and taken the best." (Broad, 1990).

To correct the flaw in the telescope, NASA devised and executed a bold in-space repair mission, carried out in 1993 by astronauts from the space shuttle *Endeavour*. The replacement system was designed to compensate permanently for the primary mirror's aberrations. This time NASA's gamble paid off, and the flawless pictures delivered by Hubble have permitted scientists to look deeper into space than ever before.

The cost of the *Endeavour* repair mission was $629 million, far in excess of the cost avoided by forgoing the land-based assembly test.

Joanna Jackson

Figure 7.4 Hubble backup mirror built by Eastman Kodak. (Used with permission from the National Air and Space Museum, Smithsonian Institution.)

Verify Requirements

We identify two steps in verifying requirements:

1. Conduct design reviews.
2. Develop a test plan.

Conduct Design Reviews

One of the most valuable techniques available to validate a design is to conduct design reviews (Karas, 2006). There are two basic types of design reviews: (1) a *customer design review,* which is a presentation to a group of individuals who collectively represent the potential or known customers and the ultimate users of the proposed product or service; and (2) an *internal design review,* which is a presentation to so-called subject matter experts within the design organization. A *subject matter expert* is someone who may not be directly associated with the project but who has deep knowledge about some aspect of design, engineering, or operation that is relevant to the project.

The purpose of the customer design review is to ensure that the product or service being built will meet the needs of the customer. It is an opportunity to capture previously unstated customer requirements and to correct misconceptions of customer needs. These new requirements will often be behavioral requirements: functions that the system must perform.

The purpose of the internal design review is to ensure that the product or service can be built, that it will meet the standards of the design organization and other regulatory bodies, and that it will live up to the mission statement of the owner. The internal design review is an opportunity to capture safety, regulatory, and other requirements that are outside the experience of the design team. These new requirements will typically be *nonbehavioral requirements*—that is, restrictions on *how* the system may perform certain functions, rather than behavioral requirements that specify *what* functions the system must perform.

Figure 7.5 displays a table listing subject matter experts in the rows and, in the columns, the types of requirements that might surface in a design review from these individuals. The shaded cells indicate that some subject matter experts may be assigned primary responsibility for identifying the nonbehavioral requirements in certain areas. For example, hardware systems often must be installed on customer sites. As indicated in the table, a subject matter expert experienced in the difficulties of installation engineering may critique a design from this perspective and generate numerous requirements that the design team may not have thought about.

Over the course of developing a complex system, many design reviews will be conducted. Some reviews will be for the subsystems, and other reviews will be for the overall design. Each design review will have a different focus, and the state of the design will be at different stages of development. Consequently, it is difficult to characterize these reviews completely. However, the following steps describe the conduct of a design review that is appropriate for the level of this textbook. Whether it is a customer design review or an internal design review, we recommend that the focus of the meeting be placed on the behavior of the system rather than its structure. The operational description templates developed in the previous chapter are therefore central to describing the system to a review panel.

1. *Determine if entry criteria are met.* It is important not to waste the time of everyone invited to a design review. Criteria need to be established and met to ensure that the design is

	Adaptation Requirements	Safety Requirements	Security and Privacy Requirements	States and Modes Requirements	Physical Characteristics Requirements	System Quality Factors	Environmental Requirements	Growth Requirements	Design and Construction Requirements	Computer Hardware Requirements	Computer Software Requirements	Computer Communication Requirements	Logistics Requirements	Quality Assurance Requirements	System Delivery Requirements	Verification and Validation Requirements
Systems Engineering	X	X	X	■	X	X	■	■	X	X	X	■	X	X	X	X
Software Engineering								X			■	X				
Hardware Engineering	X				X		X	X	■	■		X			X	
Integration and Test																■
Reliability, Maintainability, and Support					X		X	X	X				■		■	
Installation Engineering	■	X			■		X	X	X	X						
Quality						■								■		
Safety		■														
Security			■													
Human Factors				X												

■ = Primary responsibility   X = Secondary responsibility

Figure 7.5 Responsibilities for nonbehavioral requirements. (Used with permission from the Lockheed Martin Corporation.)

sufficiently mature to warrant a formal review meeting. For example, Table 7.1 lists the design documentation components specifically considered in this text. This table could be used as a checklist of the documentation required before conducting a design review.

2. *Schedule the review meeting.* If the design review is an internal review, then the list of subject matter experts must be identified and the invitations issued. Similarly, if it is a customer design review, then the organization representing the customer must provide a review panel that will adequately represent the viewpoints of customer and user. The complete set of operational description templates should be printed and distributed to the members of the review panel at least two weeks prior to the meeting.

3. *Present the design.* The design can be described at a high level using integrated product concept sketches, functional interrelationship matrices, state change matrices, and interface, link, and interaction matrices. However, in terms of eliciting feedback from a review panel,

Table 7.1 Checklist for Design Review Entry Criteria

Mission statement or operational need statement
Use case behaviors
Functional requirements
Voice of the customer and product objectives
Competitive analysis
Performance requirements
House of quality documentation
Customer value proposition
Major design alternatives
Integrated product concepts
Pugh analysis of major design decisions
Operational description templates
Functional interrelationship matrix
State change matrix
Interface, link, and interaction matrices
Test plan
Failure modes and effects analysis
Project development schedule

it is the detailed presentation of system behavior that is often the most useful. Customers who are quiet during a presentation of the system architecture often become quite animated once they see how the system is expected to perform a task. For this reason, we recommend that the owner of each operational description template (the team member who created it) provide an overview of the task flow as part of the review.

4. *Collect review panel comments.* Provide each member of the review panel with a method to submit his or her comments in writing. For example, you could provide an outline of the presentation with abundant white space in which to write comments, or you could provide a worksheet in which the reviewer formally writes additional design requirements. After the design review, collect these comments and requirements and then summarize them in a design review report.

5. *Use exit criteria to assess whether the review was passed.* The members of the review panel should be tasked with determining whether or not the design review was passed. For this, exit criteria must be established in advance and distributed to the review panel. These criteria could include the completeness of the documentation, the satisfaction of customer need, the technical viability of the design, and the acceptability of project risk.

6. *Reschedule if the review did not pass.* Failure to pass a review is not a bad event and may not at all be the fault of the design team. For example, it may be that the customer requirements are at fault. The *Vasa* shipbuilding project (Chapter 1) might have benefited from an internal design review with a subject matter expert on stability. A design review failure in such cases may save lives.

The design review is also an opportunity for a review panel to evaluate all of the collected design requirements (behavioral and nonbehavioral), and ensure that they are valid. A *valid requirement* possesses the following properties: it is

- *correct*—it specifies something that is actually required;
- *clear and precise*—it is easily understood by everyone who must work with it;
- *unambiguous*—it has only one interpretation;
- *complete*—it omits no pertinent information;
- *objective*—it leaves no room for subjective interpretation;
- *verifiable*—there is a way in which its satisfaction can be determined; and
- *consistent*—it does not contradict another requirement.

Examples of Valid and Invalid Requirements

The following isolated requirements are likely to be valid:

- The aircraft gross lift-off weight shall be less than 10,000 kg.
- The sea-level standard-day thrust shall be no less than 99,000 N.
- The launch-vehicle upper stage shall provide the spacecraft with positive detection of separation via a +5 V signal on pin 8 of the umbilical connector.

The following requirements, on the other hand, are certainly invalid:

- The robot shall be robust to failures.
- The system shall degrade gracefully.
- The navigation system shall be user friendly.

It is left as a discussion to identify the properties of valid requirements that are violated by these latter requirements.

Develop the Test Plan

Developing a test plan is often left until too late in the design cycle. If you wait until the product or system has been designed and built, then the tests you are likely to develop will be based on what you think the capabilities of the system are. That is, you will test the system based on its design rather than based on its requirements. For example, in the toy catapult project, suppose we had designed and built a trigger mechanism. Studying that particular trigger mechanism, we could design tests to make sure that it operates reliably under different projectile loads. We may have forgotten by this point that one of the originating requirements was for a mechanism that could be triggered from both the vertical and the horizontal directions. It is easy to lose sight of your original requirements.

The mark of a systems approach to design is to develop the test plan early and to base the tests on the originating requirements. Publishing the test plan early in the design cycle and setting a schedule for testing can be effective in keeping the entire design team focused on the system requirements for success.

The four steps in developing a test plan are to:

1. identify the behavioral test sequences;
2. develop the behavioral test methodology;
3. repeat for the nonbehavioral tests; and
4. map test activities to system requirements.

Identify Behavioral Test Sequences

Given all the effort that has gone into describing the behavior of the system, it is relatively easy to develop test sequences from the operational description templates (ODTs). It is a matter of supplementing the operational description with information that would be needed by, or useful to, a test engineer. These data can be inserted directly into a copy of the ODT to describe the test procedure.

Tables 7.2–7.6 show an ODT for the behavior "child plays with toy" supplemented with such test information. In the preamble to the test procedure (Table 7.2), we make note of any special lab setup requirements (goggles and gloves in this case) and list required test equipment. In the body of the behavior, Tables 7.3–7.5, we add two columns to the spreadsheet. In the first column, "expected result," we list expected results at different stages of the behavior. For example, whenever the system state is expected to change, it should be possible to observe it in the new state. We may also want to check that the timing happens as specified in the timing column. In the second column, "actual result," we leave a space for the test engineer to record the actual results. If the test is run multiple times, then we can simply add more columns to track the actual results from these different trials. In the postbehavior section, Table 7.6, we add notes to the test engineer, such as the need to test the trigger mechanism from different directions.

The remaining behavioral test procedures for the toy catapult are included in the spreadsheet file that accompanies this section.

Table 7.2 Behavioral Test Procedure

Test Procedure: Child Plays with Toy
Initial conditions
1. System is in the unloaded state
Lab setup
1. Per OR.14, test subject (adult or child) shall wear protective goggles and protective gloves
Test equipment
1. No special test equipment is required

Table 7.3 Behavioral Test Procedure

| Operator (Child) | Toy | | | | | | System State | Timing Target | Expected Result | Actual Result |
| | Energy Storage System | Projectile Containment System | Projectile Launch System | Lock and Trigger System | | Projectile | | | | |
				Locking System	Trigger System					
The child pushes the receptacle							Unloaded, stable, unarmed			
Energy transfer ("energy in")										
	The system shall store or accept external energy						Unloaded, unstable, unarmed			
The child continues pushing until the receptacle is in position **Yes** No										
Information event ("receptacle in position")										

Observe that the system detects the receptacle when it is in the proper position			Observe that the system is locked and armed	Observe that the system does not accept additional energy
1 second				
	Unloaded, locked, unarmed		Unloaded, locked, armed	
			The system shall enter the armed state	
The system shall detect the receptacle in proper position	The system shall secure its receptacle	Information event ("receptacle secured")		
			The system shall limit the energy it accepts	

Table 7.4 Behavioral Test Procedure

Child	Energy Storage	Containment	Launch	Locking	Trigger	Projectile	State	Timing	Expected	Actual
The child stops pushing										
The child loads the receptacle with projectile(s)										
Material transfer ("load")										
	The system shall accept and hold the projectile in its receptacle						Loaded, locked, armed			
						The projectile sits in the receptacle			Observe that the projectile does not fall out of the receptacle	
The child triggers the release (see note 1)										

Information event ("trigger")	The system shall detect the child's command to release (laterally or vertically)	Observe that the system detects the command to release	Information event ("release command")	The system shall release the receptacle	Loaded, released (unstable), unarmed	Observe that the receptacle is released	Information event ("release action")

1 second

Table 7.5 Behavioral Test Procedure

Child	Energy Storage	Containment	Launch	Locking	Trigger	Projectile	State	Timing	Expected	Actual
	The system shall convert energy to translational energy									
	Energy transfer ("energy out")									
			The system shall apply translational energy to receptacle					(1 second)	Observe that the receptacle moves	
			Information event ("stop")							
			The system shall stop				Loaded, stable, unarmed		Observe that the receptacle stops	
			Information event ("release contents")							

The system shall release the contents of the receptacle			Unloaded, stable, unarmed	Observe that the contents of the receptacle are ejected
Material transfer ("launch")				
	The projectile flies through the air and lands some distance away			Observe that the projectile travels at least 1 m

Table 7.6   Behavioral Test Procedure

Ending Conditions	
1.	The toy is in the unloaded state

Notes

1. This behavior must be repeated with different stimuli (child triggers release, passing toy train triggers release, falling toy triggers release)

Develop Behavioral Test Methodology

Behavioral test procedures, such as those shown in the previous section for the "child plays with toy" behavior, should be developed for each of the behaviors that were used as a source for originating requirements. A basic *test plan* is a listing of the test procedures together with a summary of the test facilities required and the entry and exit conditions for each test. An *entry condition* describes a requirement that must be satisfied before conducting the test. An *exit condition* describes a minimum requirement for passing the test.

In general, there are many different ways in which it might be acceptable to verify a requirement. Four common approaches are denoted by the letters *A, I, D, T:*

- *A (analysis).* Engineering analysis and computer simulation are often used to verify a requirement in the early stages of design. These techniques are much less expensive than physical tests. However, they typically employ mathematical approximations to the real world, and they can fail to predict behavior and physical characteristics exactly.
- *I (inspection).* Many requirements can be verified by visual inspection of the physical object.
- *D (demonstration).* Demonstrating that a concept works using a subassembly, a scale model, or a partially completed prototype is sufficient for some requirements. This is less expensive than a test requiring a full-scale working prototype.
- *T (physical test).* This refers to physical tests of a full-scale working prototype.

Table 7.7 is a test plan for the main behavioral tests of the toy catapult. The test facilities are summarized from the preamble to each test procedure (e.g., Table 7.2). In the entry condition, we note that it is not necessary to have a completed prototype for some of the tests. Test procedures TP.2, TP.3, and TP.4 could be conducted without the lock and trigger subsystem installed. This is useful information when developing a project schedule. The earlier that the tests can be performed, the sooner flaws can be discovered and corrected. Also, testing can become a bottleneck to project completion. If the testing workload can be started earlier, then the project is less likely to be delayed due to testing backlog.

Observe that we have been somewhat arbitrary in establishing exit conditions. The conditions for passing a test may require approval by the owner, the customer, or a regulatory agency. The conditions will typically require a number of repetitions of the test procedure and a test for statistical validity. It is by developing a test plan that you begin to appreciate the importance of working with clear, well-specified requirements. If you find yourself making arbitrary exit conditions, as in the case of Table 7.7, it is because you did not receive adequate guidance from the requirements

Table 7.7 Behavioral Test Plan

Test Number	Test Method	Test Facilities	Entry Condition	Exit Condition
TP.1	Test procedure: child plays with toy	Four-year-old child with protective goggles and gloves	Completed prototype	After instruction, four-year-old child is able to operate catapult. Catapult detects different trigger stimuli from both horizontal and vertical directions (five consecutive successes from each direction).
TP.2	Test procedure: child fails to arm toy		Integration of energy storage, projectile containment, and projectile launch systems	Catapult survives 50 repetitions.
TP.3	Test procedure: child releases armed toy near face of self or of another child	Apple	Integration of energy storage, projectile containment, and projectile launch systems	No bruise appears on apple 6 hours after test.
TP.4	Test procedure: child aims projectile at eyes of self or of another child	Shelled hard-boiled eggs, crayons, play scissors, blocks	Integration of energy storage, projectile containment, and projectile launch systems	No egg shows puncture to depth greater than 1 mm.
TP.5	Test procedure: child uses pet rodent as projectile	Peeled half-banana	Completed prototype	Banana survives five consecutive launches without breaking apart or deforming more than 1 cm.
TP.6	Test procedure: child drops or throws the toy	Concrete floor	Completed prototype	Toy survives six consecutive drops from height of 1 m using six orthogonal orientations.

themselves. Ideally, the exit conditions should be derived directly from the wording of the requirements. Cycling through the process of writing requirements and then developing test plans is the best way to improve your ability to write good requirements. That is, as you write requirements for a system, you should be asking yourself how each requirement can be tested and how you will judge success.

Repeat for Nonbehavioral Tests

In addition to testing the behaviors of the system, it is also necessary to verify the nonbehavioral requirements. Many of these requirements arise from the house of quality analysis (Chapter 3). Other nonbehavioral requirements can emerge from the design reviews. For each requirement, you must develop a test method.

Table 7.8 lists a test plan designed to test the nonbehavioral requirements of the toy catapult. We continue the test procedure numbering from Table 7.7. We include the statement of the originating requirement to make it easier to understand the test method. Observe that we use a mixture of approaches, including analysis (A), demonstration (D), and physical test (T). As owner of the project, we decided that some of the requirements (such as the required construction time) did not require the expense of a physical test: An engineering study predicting the result would be sufficient. Some of the entry conditions require only component samples. These tests can be performed very early in the design cycle.

We wrote one of the exit conditions (TP.7) to illustrate the use of a statistical acceptance test. In this case, it is planned that the launch velocity test will be repeated a number of times, and a specific measurement of the launch velocity will be derived from each test run. From the sample of measurements, the mean and standard deviation are calculated (spreadsheet skill: formulas for mean and standard deviation). A *statistical acceptance test* is an exit condition that the sample mean be greater than a requirement minimum (or less than a requirement maximum) by at least a given number of sample standard deviations. The so-called *Six-Sigma* methodology is based on the goal of having the sample mean lie six sample standard deviations below any upper requirement and six standard deviations above any lower requirement. This yields a total of 12 standard deviations ("sigmas").

Because the toy catapult is a hobbyist project, we have not been so aggressive in setting goals. We used a goal of three standard deviations. In particular, we require that the average launch velocity of the toy catapult for launching a ballistic ball (BB) exceed 6 mps by three sample standard deviations.

Suppose, for illustration, we made 10 runs of a prototype toy catapult and recorded the launch velocities from each run, as shown in the column labeled "first prototype" in Table 7.9. The sample average of these 10 runs is 6.45 mps. The sample standard deviation is 0.207. If we subtract three sample standard deviations from the sample mean, we get a lower limit of 5.8 mps. This is less than the requirement of 6 mps, so we declare that this prototype *fails* the test. It does not satisfy the minimum launch velocity requirement. Suppose, further, that we improve the design and construction of the prototype and try again, obtaining 10 velocity measurements as shown in the column labeled "improved prototype." In one sense, this prototype is inferior to the original prototype: The sample mean velocity is only 6.2 mps. However, the sample standard deviation is much lower. The improved prototype has a more *consistent* performance. This enables the improved prototype to pass the test. The mean launch velocity minus three sample standard deviations is 6.1 mps, which satisfies the design requirement.

Table 7.8   Test Methodologies for Nonbehavioral Requirements

Requirement No.	Requirement	Abstract Name	Test No.	Test Method	Verification Method (A, I, D, T)	Test Facilities	Entry Condition	Exit Condition
OR.14	The maximum launch velocity of the toy catapult shall be at least 6 mps.	Maximum launch velocity	TP.7	Measure distance thrown of a small ballistic ball (BB). Derive launch velocity from analysis, neglecting air resistance. Repeat at least 10 times.	D	Flat floor, ballistic ball (BB), measuring tape	Integration of energy storage, projectile containment, and projectile launch systems	Derived velocity exceeds requirement by three standard deviations of sample.
OR.15	The maximum mass launch capability of the toy catapult shall be at least 150 g.	Maximum mass launch capability	TP.8	Measure distance thrown of a projectile with mass of 150 g. Repeat at least five times.	D	Projectile with mass of 150 g, measuring tape	Integration of energy storage, projectile containment, and projectile launch systems	Average distance thrown exceeds 0 cm in five consecutive trials.
OR.16	The storage volume of the toy catapult shall not exceed 4,500 cc.	Storage volume	TP.9	Calculate from design drawings.	A	None	Detailed design drawings	Calculated volume satisfies requirement.

(Continued)

Table 7.8 Test Methodologies for Nonbehavioral Requirements (Continued)

Requirement No.	Requirement	Abstract Name	Test No.	Test Method	Verification Method (A, I, D, T)	Test Facilities	Entry Condition	Exit Condition
OR.17	The mean cycles to failure of the toy catapult shall be at least 1,000.	Mean cycles to failure	TP.10	Record history of field usage of prototype.	T	Log sheet to track number of launches	Completed prototype	Prototype survives 1,500 launches.
OR.18	The breaking force of a critical component such as the crossbar shall be at least 200 N.	Breaking force of critical components	TP.11	Support crossbar (or other critical component) at ends. Suspend a weight of mass (200 N)/(9.8 m/s^2) = 20.4 kg from center of crossbar.	D	Weight of mass 20.4 kg, supports, string.	Component sample	Component supports load for 1 hour without breaking.
OR.19	The cost of a replacement for a critical component such as a crossbar shall not exceed $4.	Repair part cost	TP.12	Calculate from design drawings.	A	None	Detailed design drawings	Calculated cost satisfies requirement.

OR.20	If a critical component such as a crossbar breaks and is repaired using typical home-applied adhesive, the tensile strength of home-applied adhesive bond shall be at least 15 MPa	Tensile strength of home-applied adhesive bond	TP.13	Break crossbar and repeat TP.11 with repaired crossbar. Use household glue appropriate for material (wood or plastic)	D	Same setup as TP.11 with addition of household glue	Component sample	Repaired component supports load for 1 hour without breaking
OR.21	The time required for disassembly and reassembly shall not exceed 120 minutes	Time for disassembly and reassembly	TP.14	Estimate from process plan	A	None	Detailed design drawings and process plan	Estimated time satisfies requirement

(Continued)

Table 7.8 Test Methodologies for Nonbehavioral Requirements (Continued)

Requirement No.	Requirement	Abstract Name	Test No.	Test Method	Verification Method (A, I, D, T)	Test Facilities	Entry Condition	Exit Condition
OR.22	The material cost of the toy catapult shall not exceed $20	Material cost	TP.15	Calculate from design drawings and vendor quotations	A	None	Detailed design drawings	Calculated cost satisfies requirement
OR.23	The time to manufacture the toy catapult shall not exceed 60 minutes per toy	Time to manufacture	TP.16	Estimate from process plan	A	None	Detailed design drawings and process plan	Estimated time satisfies requirement

Table 7.9   Statistical Acceptance Test for Maximum Launch Velocity

Run Number	First Prototype	Improved Prototype
1	6.45	6.15
2	6.53	6.21
3	6.08	6.20
4	6.61	6.19
5	6.36	6.22
6	6.24	6.16
7	6.56	6.16
8	6.68	6.21
9	6.30	6.20
10	6.73	6.23
Sample mean	6.45	6.19
Sample std. dev.	0.21	0.03
Test	5.83	6.11
Test acceptance	Fail	Pass

Map Test Activities to System Requirements (VCRM)

Once a test plan has been developed, it is important to check if it is complete. The process for checking completeness is to map each test activity to the requirement it is intended to verify. If the test verifies a derived requirement, rather than an originating requirement, then continue the process by tracing the derived requirement back to an originating requirement. In this way, it should be possible to trace each test back to an originating requirement.

Table 7.10 checks each derived functional requirement for the toy catapult and maps it to a test procedure as well as back to an originating requirement. Table 7.11 traces test procedures associated with nonbehavioral requirements to those requirements.

To check for completeness of the test plan, it is useful to construct a *verification cross-reference matrix* (VCRM), which is a matrix that maps test procedures to originating requirements. An "x" in the matrix indicates that a particular test procedure is used to verify a particular originating requirement.

Table 7.12 is a VCRM for the toy catapult. From this matrix, we discover that OR.10 ("rodent escapes") has no test procedure (there are no x's in this column) and test TP.2 ("child fails to arm toy") has no originating requirement (there are no x's in this row). These are valuable discoveries that may cause us to extend the test plan or modify the originating requirements. OR.13 (no harm) has no test procedure, but it is satisfied by the test design. OR.6 and OR.7 (labeling) are to be satisfied with packaging material design. We have decided to postpone packaging design to a later design cycle.

Table 7.10   Trace Test Procedures to Originating Requirements

Derived From	Index	Derived Functional Requirement	Function Name	Test Procedure
OR.5	DR.1	The system shall store or accept external energy.	Store energy	TP.1
OR.9	DR.2	The system shall limit the energy it accepts.	Limit energy	TP.1
OR.5	DR.3	The system shall convert energy to translational energy.	Convert energy	TP.1
OR.5	DR.4	The system shall release stored energy.	Release energy	TP.1
OR.1	DR.5	The system shall detect the receptacle in proper position.	Detect arming position	TP.1
OR.2	DR.6	The system shall secure its receptacle.	Secure receptacle	TP.1
OR.5	DR.7	The system shall release the receptacle.	Release receptacle	TP.1
OR.3	DR.8	The system shall accept and hold the projectile in its receptacle.	Hold projectile	TP.1
OR.5	DR.9	The system shall release the contents of the receptacle.	Release projectile	TP.1
OR.2	DR.10	The system shall enter the armed state.	Enter armed state	TP.1
OR.4	DR.11	The system shall detect the command to release (laterally or vertically) from the child.	Detect release command	TP.1
OR.5	DR.12	The system shall apply translational energy to receptacle.	Apply energy	TP.1
OR.5	DR.13	The system shall stop.	Stop	TP.1
OR.8	DR.14	The moving parts of the system shall be prevented from striking the face of a child with sufficient force to cause a bruise.	Do not hurt child's face	TP.3
OR.9	DR.15	The system shall not launch a blunt projectile with sufficient force to penetrate a child's eye.	Do not hurt child's eye	TP.4
OR.9	DR.16	The system shall not launch a rodent with sufficient force to injure the rodent when it lands.	Do not injure rodent	TP.5
OR.11	DR.17	The system shall successfully complete each cycle of use.	Survive repetition	
OR.12	DR.18	The system shall survive a collision with a hard surface in working order.	Survive collision	TP.6

Table 7.11 Trace Test Procedures to Nonbehavioral Requirements

Requirement No.	Nonbehavioral Requirement	Abstract Name	Test Procedure
OR.14	The maximum launch velocity of the toy catapult shall be at least 6 mps.	Maximum launch velocity	TP.7
OR.15	The maximum mass launch capability of the toy catapult shall be at least 150 g.	Maximum mass launch capability	TP.8
OR.16	The storage volume of the toy catapult shall not exceed 4,500 cc.	Storage volume	TP.9
OR.17	The mean cycles to failure of the toy catapult shall be at least 1,000.	Mean cycles to failure	TP.10
OR.18	The breaking force of a critical component such as the crossbar shall be at least 200 N.	Breaking force of critical components	TP.11
OR.19	The cost of a replacement for a critical component such as a crossbar shall not exceed $4.	Repair part cost	TP.12
OR.20	If a critical component such as a crossbar breaks and is repaired using typical home-applied adhesive, the tensile strength of the home-applied adhesive bond shall be at least 15 MPa.	Tensile strength of home-applied adhesive bond	TP.13
OR.21	The time required for disassembly and reassembly shall not exceed 120 minutes.	Time for disassembly and reassembly	TP.14
OR.22	The material cost of the toy catapult shall not exceed $20.	Material cost	TP.15
OR.23	The time to manufacture the toy catapult shall not exceed 60 minutes per toy.	Time to manufacture	TP.16

238 Getting Design Right: A Systems Approach

Table 7.12 Verification Cross-Reference Matrix (VCRM) for Toy Catapult

Test (row) Is Used to Verify Originating Requirement (column)	OR.1	OR.2	OR.3	OR.4	OR.5	OR.6	OR.7	OR.8	OR.9	OR.10	OR.12	OR.13	OR.14	OR.15	OR.16	OR.17	OR.18	OR.19	OR.20	OR.21	OR.22	OR.23
TP.1	×	×	×	×	×				×													
TP.2																						
TP.3								×														
TP.4									×													
TP.5									×													
TP.6											×											
TP.7													×									
TP.8														×								
TP.9															×							
TP.10																×						
TP.11																	×					
TP.12																		×				
TP.13																			×			
TP.14																				×		
TP.15																					×	
TP.16																						×

The test plans (Tables 7.7 and 7.8) and the VCRM (Table 7.12) are the major outputs of this planning process.

Manage Risks

> For want of a nail, the shoe was lost,
> For want of the shoe, the horse was lost,
> For want of the horse, the rider was lost,
> For want of the rider, the message was lost,
> For want of the message, the battle was lost,
> For want of the battle, the war was lost,
> For want of the war, the kingdom was lost,
> And all for the want of a shoe!

Anonymous

Risk is the combination of chance and a negative outcome. Being struck by lightning is a frightening prospect (a negative outcome), but, fortunately for most of us, the chance of getting hit by lightning is typically so remote that it is not a risk to which we give much thought in our daily lives. When circumstances change, however, such as finding ourselves on a golf course in a thunderstorm, the risk increases (because the chance has increased), and we are wise to take action to reduce the risk.

Design risk is the possibility that the product or system being designed may fail in some way that degrades or nullifies the mission or, worse, causes hardship or injury to people. *Design risk management* is the activity of proactively identifying risks and attempting to prevent or minimize the more serious ones. An action taken to reduce a risk is called a *corrective action*. In general, a corrective action attempts to reduce the chance of the negative outcome (for example, by running to the clubhouse or into a gully to avoid lightning) or to reduce the severity of the negative outcome (for example, by carrying a first aid kit so that rapid treatment can be given to anyone in your party who is injured by lightning). A collection of corrective actions designed to deal with a risk is called a *risk mitigation strategy*.

We can learn a lot about risk by studying past design failures. Henry Petroski (1992) studied famous failures in civil engineering, such as the collapse of the Tacoma Narrows bridge. *Root cause analysis* is the attempt to isolate the most fundamental factor or combination of factors that led to a particular failure. The explosion of the space shuttle *Challenger* in 1986, for example, was traced to the failure of an O-ring seal in its right solid rocket booster. Root cause analysis frequently reveals very simple causes for catastrophic failures. The nursery rhyme at the beginning of this section ("For the want of a nail...") captures the flavor of many of these findings.

We have attempted to deal with design risk throughout this book. Recall that the purpose of secondary use cases was to identify unusual and potentially harmful user behaviors. Concerns for safety and reliability guided the setting of technical performance measures in the house of quality approach. Also, we estimated the reliability of the system by adding up the estimated failure rates of all of its components. These techniques do not exhaust the systematic ways in which a design can be studied for its potential for failure and harm. In this section we consider one more technique of design risk management that has been in use for decades. *Failure modes and effects analysis* (FMEA) is a formal technique to identify potential failures of a design, and to implement corrective actions while the product is still in the development stage (NASA). It was originally introduced by the U.S. military in 1949, and it was used for the *Apollo* space missions in the 1960s.

It was adapted for use in the automotive industry in the 1980s by the Ford Motor Company after the discovery of a life-threatening design flaw in the fuel tank of the Ford Pinto. It is now used widely in many industries.

Failure modes and effects analysis is a bottom-up approach to risk management. It is a systematic look at the design from a low level (the subsystem or component levels), asking how a single failure at this level can affect the system as a whole. A top-down approach to risk management is *fault tree analysis* (FTA), in which one starts with a potential major system failure, and investigates all the different ways in which this failure could occur and the combination of factors that must be present in each occurrence. The calculations in fault tree analysis are outside the scope of this book. We focus on the simpler but still highly valuable technique of FMEA.

Conduct Failure Modes and Effects Analysis

There are nine steps in conducting an FMEA:

1. Select items or functions for analysis.
2. Identify failure modes for each item.
3. Assess the potential impact of each failure mode.
4. Brainstorm possible causes for each failure mode.
5. Suggest corrective actions for each possible cause.
6. Rate the severity of the potential impact.
7. Rate the likelihood of occurrence of each possible cause.
8. Assess the risk.
9. Prioritize the corrective actions.

Select Functions

As mentioned, FMEA is a bottom-up approach to risk management. One typically starts at a very low level in the design, such as the component level, and investigates the risk of failure for each item at this level. FMEA, however, is not restricted to the component level. We can apply it to the toy catapult design even though we are still at the subsystem design level (where Chapter 6 ended). Table 7.13 lists the subsystems that have been identified for the toy catapult. In deciding what to focus on, you can think in terms of items, such as the subsystems we have identified, or in terms of functions that the system must perform. Because each of our subsystems for the toy catapult has

Table 7.13   Select Items or Functions for Bottom-Up Failure Analysis

Identification of Item or Function
Projectile launch system
Energy storage system
Projectile containment system
Lock and trigger system
Body (enclosure system)

Table 7.14   Identify Failure Modes at This Level

Failure Mode No.	Identification of Item or Function	Failure Mode
F.1	Projectile launch system	Failure to stop
F.2		Failure to transfer energy to receptacle
F.3	Energy storage system	Failure to capture energy
F.4	Projectile containment system	Attachment failure
F.5		Failure to contain projectile
F.6		Failure to release projectile
F.7	Lock and trigger system	Failure to lock
F.8		Failure to arm
F.9		Failure to release
F.10		Delayed release
F.11	Body (enclosure system)	Failure to enclose moving parts

functional requirements, it does not matter whether we think in terms of subsystems or functions: We are likely to discover the same set of risks with either approach.

Identify Failure Modes

The next step in FMEA is to identify the *failure modes*—that is, the ways in which the item or function could fail. These are typically found by reflecting on the functional requirements of the item and listing the negation of each of these functions. Table 7.14 lists the failure modes for the toy catapult subsystems. Note that each failure mode is assigned a number and that one subsystem can experience many failure modes. Keep the descriptions of the failure modes concise.

Assess Potential Impact of Failure

The next step in FMEA is to imagine the failure and ask what negative outcomes could be associated with such a failure. NASA suggests thinking about negative outcomes at three levels:

1. Local. This would be the immediate effect. For example, an O-ring failure permits gas to escape.
2. System. For example, escaping gas ignites and causes the rocket booster to explode.
3. Mission. For example, the space flight terminates, and the crew is killed.

In Table 7.15, we have attempted to apply this thinking to the failure modes identified in Table 7.14. Two discoveries emerge from this. The first is that by picturing the receptacle coming loose and flying off the catapult, we identified a potential source of injury to the child that our previous approaches did not reveal. Similarly, the purpose of the enclosure system is to protect

Table 7.15   Assess Impact of Failure at This and Higher Levels

Failure Mode No.	Identification of Item or Function	Failure Mode	Failure Effects (a. Local; b. System; c. Mission)
F.1	Projectile launch system	Failure to stop	b. Projectile not launched or launched late; c. Failed or degraded mission
F.2		Failure to transfer energy to receptacle	b. Projectile launched with little or no force; c. Failed or degraded mission
F.3	Energy storage system	Failure to capture energy	b. Unable to launch projectile; c. Failed mission
F.4	Projectile containment system	Attachment failure	b. Receptacle and projectile launched; c. Injury to child or bystander
F.5		Failure to contain projectile	b. Failure to launch; c. Failed mission
F.6		Failure to release projectile	b. Failure to launch at optimum point; c. Degraded mission
F.7	Lock and trigger system	Failure to lock	b. Unable to launch projectile; c. Failed mission
F.8		Failure to arm	b. Unable to launch projectile; c. Failed mission
F.9		Failure to release	b. Unable to launch projectile; c. Failed mission
F.10		Delayed release	b. Unexpected launch; c. Injury to child or bystander
F.11	Body (enclosure system)	Failure to enclose moving parts	b. Exposure of moving parts; c. Injury to child's fingers or face

the child's face from being hit by moving parts. However, the enclosure system could create an opportunity for the child's fingers to be pinched. It has taken until now in the design process for these concerns to surface. This is precisely the purpose of FMEA: to catch such potential negative outcomes at a stage when the design can still be modified to correct the problem.

Brainstorm Possible Causes

The next step in FMEA is to identify possible causes for the failure modes. This is like conducting a root cause analysis, but it is in response to an imagined failure, rather than to an actual one. One of the techniques of root cause analysis that is also useful in this context is to list categories and then consider how failures in each of those categories might result in the particular failure mode. Four frequently used categories are "the 4 Ms":

1. "Man" suggests various types of human error such as lack of experience, operator inattention, and lack of training.
2. "Machine" suggests failures such as equipment breakdown and improper maintenance or setup.
3. "Method" includes software errors, poor documentation, incorrect operating procedures, and poor inspection and testing methods.
4. "Material" includes substandard material or components, metal fatigue, or excessive stress forces.

Table 7.16 lists possible causes for the failure modes of the toy catapult. Note that there can be multiple possible causes for the same failure mode. All of the causes listed in Table 7.16 would be classified in the machine or material categories. We did not explore possible causes in the other categories. You can likely think of many more possible causes.

As a designer or an engineer, once you think of how to solve a problem (such as using a weighted pendulum for the energy storage system), you are ready to move on and solve other problems. FMEA is interesting because it forces you to think about how your solution may not always work. How could a weighted pendulum "fail to capture energy" (failure mode F.3)? It would have to be blocked in some way. But that should be easy to fix: just remove the blockage. Well, if all the moving parts are enclosed in the body, then it might not be easy to remove the blockage. But if the moving parts are enclosed, then how could anything enter the body to become blockage in the first place? The answer is that children have a remarkable ability to stuff things into small places. It would be very difficult to guarantee that nothing could enter the body.

We conclude that our design concept of enclosing the moving parts of the toy for safety includes the real possibility that the child could stuff a marble or some other small object into the body that could then become jammed in the energy storage system, causing it to fail. We now look at the design of the enclosure system from a new perspective. It was introduced as a safety feature, but it could make the toy more susceptible to failure. You can expect to make many such discoveries through the FMEA process.

Suggest Corrective Actions

Once the possible causes of failure modes have been identified, the FMEA process requires you to engage in problem solving. What are the most cost-effective ways of reducing the risk posed by each possible cause? As mentioned above, corrective action can be developed either to reduce the likelihood of the negative outcome, or to reduce the severity of the negative outcome if it does occur. Thinking along one or both of these two lines of thought is helpful in discovering solutions. Another dimension to guide your thinking is to think about the timing of the corrective action. Three possible timings are

1. *Design.* What can be done to the fundamental design of the system to reduce the risk associated with this possible cause?
2. *Manufacturing process.* What improvements in the manufacturing process can reduce the risk?
3. *Operation.* Are there operational procedures to be performed by the responsible user that would reduce the risk?

Table 7.16 Brainstorm Possible Causes

Failure Mode Number	Identification of Item or Function	Failure Mode	Failure Effects (a. Local; b. System; c. Mission)	Possible Cause
F.1	Projectile launch system	Failure to stop	b. Projectile not launched or launched late; c. Failed or degraded mission	Crossbar breaks
				Crossbar shifts location
F.2		Failure to transfer energy to receptacle	b. Projectile launched with little or no force; c. Failed or degraded mission	Catapult arm breaks
				Pivot becomes sticky
F.3	Energy storage system	Failure to capture energy	b. Unable to launch projectile; c. Failed mission	Foreign object is trapped in system and causing blockage
				Disconnected (e.g., weight falls off arm)
F.4	Projectile containment system	Attachment failure	b. Receptacle and projectile launched; c. Injury to child or bystander	Glue bond between receptacle and catapult arm weakens and fails
F.5		Failure to contain projectile	b. Failure to launch; c. Failed mission	Receptacle breaks
F.6		Failure to release projectile	b. Failure to launch at optimum point; c. Degraded mission	Shape of receptacle is deformed
F.7	Lock and trigger system	Failure to lock	b. Unable to launch projectile; c. Failed mission	Worn connection point
F.8		Failure to arm	b. Unable to launch projectile; c. Failed mission	Removable pin is lost
F.9		Failure to release	b. Unable to launch projectile; c. Failed mission	Foreign object is trapped in system and causing blockage
F.10		Delayed release	b. Unexpected launch; c. Injury to child or bystander	Latch mechanism becomes sticky
F.11	Body (enclosure system)	Failure to enclose moving parts	b. Exposure of moving parts; c. Injury to child's fingers or face	Fasteners fail

Table 7.17 lists corrective actions developed to address the possible causes of failure for the toy catapult. In some cases, such as risk F.4, a risk mitigation strategy consisting of several actions is proposed. Most of these actions would be classified as reducing the likelihood of the possible cause occurring. However, in the case of a foreign object causing blockage in the energy storage system (F.4), the recommended action is to change the design so that at least one side of the enclosure is removable. This, obviously, will not reduce the likelihood of a foreign object entering the body of the toy and causing blockage. However, it will permit the parent to fix the problem more easily by giving adults easy access to the interior of the toy. In this way, the loss of function is only temporary. It is a corrective action designed to reduce the severity of the negative outcome.

Note that the actions have been classified as changes to design, manufacturing process, or operation, as suggested before. An example of a manufacturing process change is in response to attachment failure (F.4) caused possibly by a failure of glue to hold the receptacle to the catapult arm. The quality of the glue bond could be improved by better preparing the surfaces (sanding the wood) prior to application of the glue. An example of an operational change is to provide the operator (the parent) with instructions on how to clean the lock and trigger mechanism if it becomes sticky.

Identifying corrective actions is not the same as implementing them. The corrective actions in Table 7.17 are not required changes. A cost-benefit analysis must be conducted to decide which changes to implement. We now enter a judgment phase of the FMEA process to decide on whether the benefits of risk reduction are worth the cost. For this, it is useful to develop a quantitative approach to measuring risk. The next steps describe that approach.

Rate the Severity of Impact

The quantitative approach to risk assessment requires that we measure the severity of the negative outcome separately from the likelihood of its occurrence. For this, we need some scale against which to measure severity. Different kinds of risks may require different severity scales. Table 7.18 displays a severity scale that is used frequently in industry (aerospace and automotive). Large projects such as space missions or public civil engineering construction projects have greater risks associated with them, and they require an expanded scale. Simpler scales of severity can also be found. The scale in Table 7.17 is more than adequate for the toy catapult project.

Having established a severity rating system, the next step is to assess the severity of the different risks against this scale. Table 7.20 shows the completed FMEA worksheet in which we have included our judgment of the severity associated with each negative outcome.

Rate the Likelihood of Causal Occurrence

Similar to the process of assessing severity, we next establish a rating system to assess the likelihood of occurrence of a possible cause. Table 7.19 is an occurrence scale used frequently in industry. The "user's required time" is taken to be the period of service during which the user is likely to use the system, measured in hours or days of actual operation. MTTF is the concept of mean time to failure that was introduced in Chapter 6. It is applied to the particular component or subsystem that is featured in the possible cause.

Having established an occurrence rating system, the next step is to assess the occurrence of the different possible causes against this scale. Table 7.20 also shows our judgment of the likelihood of occurrence associated with each possible cause.

Table 7.17 Suggest Corrective Action

Failure Mode No.	Identification of Item or Function	Failure Mode	Failure Effects (a. Local; b. System; c. Mission)	Possible Cause	Corrective Action (a. Design; b. Manufacturing Process; c. Operation)
F.1	Projectile launch system	Failure to stop	b. Projectile not launched or launched late; c. Failed or degraded mission	Crossbar breaks	a. Use stronger material
				Crossbar shifts location	a. Use enclosure to reinforce crossbar position
F.2		Failure to transfer energy to receptacle	b. Projectile launched with little or no force; c. Failed or degraded mission	Catapult arm breaks	a. Use stronger material
				Pivot becomes sticky	a. Use metal or plastic liner
F.3	Energy storage system	Failure to capture energy	b. Unable to launch projectile; c. Failed mission	Foreign object trapped in system is causing blockage	a. Design at least one side of enclosure to be removable
				Disconnected (e.g., weight falls off arm)	a. Redesign connection
F.4	Projectile containment system	Attachment failure	b. Receptacle and projectile launched; c. Injury to child or bystander	Glue bond between receptacle and catapult arm weakens and fails	a. Use fastener as well as glue. b. Improve surface preparation step
F.5		Failure to contain projectile	b. Failure to launch; c. Failed mission	Receptacle breaks	a. Use stronger material

F.6		Failure to release projectile	b. Failure to launch at optimum point; c. Degraded mission	Shape of receptacle is deformed	a. Use rigid material
F.7	Lock and trigger system	Failure to lock	b. Unable to launch projectile; c. Failed mission	Worn connection point	a. Cover connection point with metal band
F.8		Failure to arm	b. Unable to launch projectile; c. Failed mission	Removable pin is lost	a. Secure pin with more reliable chain or string
F.9		Failure to release	b. Unable to launch projectile; c. Failed mission	Foreign object trapped in system is causing blockage	a. Design at least one side of enclosure to be removable
F.10		Delayed release	b. Unexpected launch; c. Injury to child or bystander	Latch mechanism becomes sticky	a. Design at least one side of enclosure to be removable; c. Include cleaning directions
F.11	Body (enclosure system)	Failure to enclose moving parts	b. Exposure of moving parts; c. Injury to child's fingers or face	Fasteners fail	a. Use redundant number of fasteners (more screws than needed)

Table 7.18   Establish a Rating System for Severity

Severity	Description	Explanation
1	None	Variation within performance limits
2	Very minor	Variation correctible during production
3	Minor	Reparable within 10 minutes
4	Very low	Reparable within 30 minutes
5	Low	Reparable within 1 hour
6	Moderate	Reparable within 4 hours
7	High	Not worth repair; degraded functionality or capability
8	Very high	Not worth repair; mission failure
9	Hazardous—with warning	Affects safety of operator and others but with advance warning
10	Hazardous—without warning	Affects safety of operator and others with no advance warning

Source: Adapted with permission from SAE J1739, p. 32, SAE International.

Table 7.19   Establish a Rating System for Likelihood of Occurrence

Occurrence	Description
1	MTTF is 50 times greater than user's required time
2	MTTF is 20 times greater than user's required time
3	MTTF is 10 times greater than user's required time
4	MTTF is six times greater than user's required time
5	MTTF is four times greater than user's required time
6	MTTF is two times greater than user's required time
7	MTTF is equal to user's required time
8	MTTF is 60% of user's required time
9	MTTF is 30% of user's required time
10	MTTF is 10% of user's required time

Source: Adapted with permission from SAE J1739, p. 34, SAE International.

Table 7.20 Completed FMEA Worksheet

Failure Mode No.	Identification of Item or Function	Failure Mode	Failure Effects (a. Local; b. System; c. Mission)	Possible Cause	Corrective Action (a. Design; b. Manufacturing Process; c. Operation)	Failure Effect's Severity	Occurrence Likelihood (Under Current Design)	Risk Priority No. (= Severity* Occurrence)	Criticality (Corrective Action Priority)
F.1	Projectile launch system	Failure to stop	b. Projectile not launched or launched late; c. Failed or degraded mission	Crossbar breaks	a. Use stronger material	6	4	24	Low
				Crossbar shifts location	a. Use enclosure to reinforce crossbar position	7	4	28	Low
F.2		Failure to transfer energy to receptacle	b. Projectile launched with little or no force; c. Failed or degraded mission	Catapult arm breaks	a. Use stronger material	6	4	24	Low
				Pivot becomes sticky	a. Use metal or plastic liner	7	3	21	Low
F.3	Energy storage system	Failure to capture energy	b. Unable to launch projectile; c. Failed mission	Foreign object trapped in system is causing blockage	a. Design at least one side of enclosure to be removable	8	8	64	Critical
				Disconnected (e.g., weight falls off arm)	a. Redesign connection	8	4	32	Intermediate

(Continued)

Table 7.20 Completed FMEA Worksheet (Continued)

Failure Mode No.	Identification of Item or Function	Failure Mode	Failure Effects (a. Local; b. System; c. Mission)	Possible Cause	Corrective Action (a. Design; b. Manufacturing Process; c. Operation)	Failure Effect's Severity	Occurrence Likelihood (Under Current Design)	Risk Priority No. (= Severity* Occurrence)	Criticality (Corrective Action Priority)
F.4	Projectile containment system	Attachment failure	b. Receptacle and projectile launched; c. Injury to child or bystander	Glue bond between receptacle and catapult arm weakens and fails	a. Use fastener as well as glue; b. Improve surface preparation step	10	7	70	Critical
F.5		Failure to contain projectile	b. Failure to launch; c. Failed mission	Receptacle breaks	a. Use stronger material	8	3	24	Low
F.6		Failure to release projectile	b. Failure to launch at optimum point; c. Degraded mission	Shape of receptacle is deformed	a. Use rigid material	7	3	21	Low
F.7	Lock and trigger system	Failure to lock	b. Unable to launch projectile; c. Failed mission	Worn connection point	a. Cover connection point with metal band	8	6	48	Intermediate

		Failure mode	Effects	Cause	Recommended actions				
F.8		Failure to arm	b. Unable to launch projectile; c. Failed mission	Removable pin is lost	a. Secure pin with more reliable chain or string	6	6	36	Low
F.9		Failure to release	b. Unable to launch projectile; c. Failed mission	Foreign object trapped in system is causing blockage	a. Design at least one side of enclosure to be removable	8	8	64	Critical
F.10		Delayed release	b. Unexpected launch; c. Injury to child or bystander	Latch mechanism becomes sticky	a. Design at least one side of enclosure to be removable; c. Include cleaning directions	10	4	40	Critical
F.11	Body (enclosure system)	Failure to enclose moving parts	b. Exposure of moving parts c. Injury to child's fingers or face	Fasteners fail	a. Use redundant number of fasteners (more screws than needed)	10	2	20	Critical

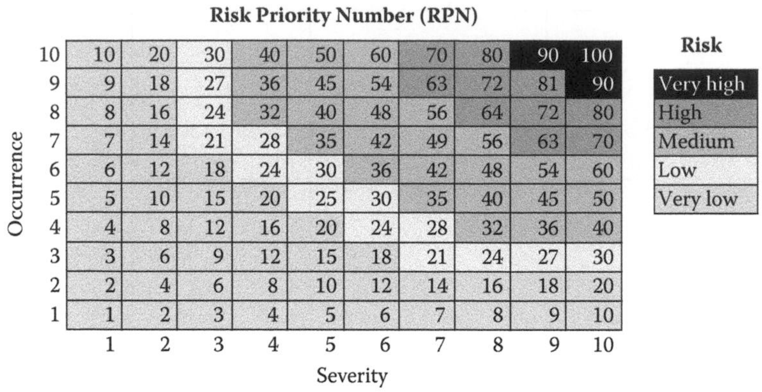

Risk Priority Number (RPN)

Occurrence											Risk
10	10	20	30	40	50	60	70	80	90	100	**Very high**
9	9	18	27	36	45	54	63	72	81	90	High
8	8	16	24	32	40	48	56	64	72	80	Medium
7	7	14	21	28	35	42	49	56	63	70	Low
6	6	12	18	24	30	36	42	48	54	60	Very low
5	5	10	15	20	25	30	35	40	45	50	
4	4	8	12	16	20	24	28	32	36	40	
3	3	6	9	12	15	18	21	24	27	30	
2	2	4	6	8	10	12	14	16	18	20	
1	1	2	3	4	5	6	7	8	9	10	
	1	2	3	4	5	6	7	8	9	10	

Severity

Figure 7.6   Measuring risk using the risk priority number.

Assess the Risk

Risk is the combination of chance and negative outcome. It is thus a two-dimensional concept.* A quantitative approach to risk is to multiply the severity rating by the occurrence rating to yield a single index, called the *risk priority number* (RPN):

$$RPN = severity\ rating\ *\ occurrence\ rating.$$

Figure 7.6 displays the risk priority numbers for different combinations of severity rating and occurrence ratings. We have included a subjective judgment of risk by shading the different RPNs.

Observe that a severe negative outcome (severity = 10) can be judged to be low risk if the occurrence is sufficiently rare.

Observe that we rate the severity of the outcome separately from the occurrence of the possible cause. The same outcome, therefore, can have different risks associated with it, depending on the different possible causes. In Table 7.20, we have calculated the RPN for each combination of outcome severity and possible cause occurrence in the toy catapult example.

Prioritize the Action

The final step in the FMEA process is to assign a priority for implementation of each corrective action. It is understood that low-priority actions may not be implemented. Critical actions are ones that should be implemented as soon as possible. You might think that it would be possible simply to set cutoff points according to risk priority numbers and then use those cutoffs to determine criticality. The problem with this approach is that if someone wanted to avoid taking action, he or she might just "juggle the numbers" to get the RPNs to fall below the cutoffs. The best course of action for a design project manager is to make sure that some corrective actions are taken. You may use the RPNs as a guide, but you might also be wise simply to work on reducing the risk associated with the most severe outcomes.

Table 7.20 shows how we have prioritized the corrective actions for the toy catapult. In general, these priorities ("critical," "intermediate," and "low") follow the RPNs, but there are some

---

* A more advanced form of FMEA introduces another dimension—namely, "detection"—under the assumption that an undetected failure can be more serious than a failure that can be detected early and for which timely correction is possible.

exceptions. We have chosen to give high priority to items that affect child safety, even if the likelihood of occurrence is low. The RPNs in this case do not really capture our assessment of risk.

Table 7.20 is the completed FMEA worksheet for the toy catapult. The FMEA approach is an extremely important tool for design risk management. It is a systematic technique for discovering and correcting potential design flaws. As a final note, observe that because the failures can all be traced to functional requirements, the corrective actions arising out of this FMEA process can be rewritten as derived requirements.

Summary

In this chapter we have considered the role of the design project manager and focused on two managerial responsibilities: verifying requirements and managing risks. We have collected these ideas under the general topic of validating the design.

To ensure that requirements are valid and that the proposed design will meet the needs of the customer and fulfill the mission statement of the owner, we recommended the use of design reviews. Customer design reviews focus on the customer and user needs. Internal design reviews subject the design to critique by subject matter experts. Design reviews are useful for discovering previously unknown requirements. For verifying requirements, we also recommended creating a test plan early in the design process. The plan should detail the test methods, test facilities, entry conditions, and exit conditions. We emphasized that it is the mark of a systems approach to base the test plan on the system requirements, rather than basing it on a particular design idea. We introduced the notion of a statistical acceptance test and explained the concept of Six Sigma quality. We also recommended constructing a verification cross-reference matrix to check whether every requirement has a way of being verified.

For managing risk, we recommended the technique of failure modes and effects analysis as a systematic way of anticipating risks, uncovering potential design flaws, and developing risk mitigation strategies. We emphasized that risk is the combination of a negative outcome and the probability of that outcome occurring. Risk mitigation strategies combine corrective actions to reduce risk by reducing the severity of the outcome and/or the likelihood of its causal occurrence. Judgment must be used to decide which corrective actions are worth implementing.

Discussion

1. Debate the wisdom of the gamble NASA took in not paying for a ground test of the assembled Hubble telescope. Use the language of failure modes and effects analysis. At least one individual should defend the NASA decision.
2. Discuss the validity of the requirements in the section entitled "Examples of Valid and Invalid Requirements."

Exercises

1. On a system of your choice, such as the BathBot, for which you have at least two behavioral requirements and two nonbehavioral requirements, develop a test plan similar to Tables 7.7 and 7.8. Only **four** tests are required. **At least one** of the tests should use a statistical acceptance test for its exit criterion.

2. On a system of your choice, complete an FMEA worksheet similar to Table 7.20 for a *single* failure mode. There should be at least two possible causes identified for this failure mode. Justify your severity and occurrence likelihood ratings and your criticality judgment in a short supplementary note.

References

Associated Press. 1990. Losing bid offered 2 tests on Hubble. *New York Times*, July 28.

Broad, W. J. 1990. Hubble has backup mirror, unused. *New York Times*, July 18.

Gradous, Lori I. As quoted in Maier and Rechtin, 1997.

Jackson, Joanna. 2009. Personal communication.

Karas, L. 2006. Owego Systems engineering principles and practices course. Lockheed Martin Systems Integration–Owego.

Leary, W. E. 1990. Senator says a bidder urged test of telescope. *New York Times*, July 7.

Maier, M. W., and E. Rechtin. 1997. *The art of systems architecting*. New York: CRC Press.

NASA. Performing a failure mode and effects analysis. Flight assurance procedure P-302-720. Goddard Space Flight Center.

Petroski, H. 1992. *To engineer is human*. New York: Vintage Books of Random House, Inc.

Society of Automotive Engineers. 2002. *Potential failure mode and effects analysis in design (design FMEA), potential failure mode and effects analysis in manufacturing and assembly processes (process FMEA), and potential failure mode and effects analysis for machinery (machinery FMEA)*. SAE J1739. SAE International.

U.S. History Encyclopedia. The Hubble Space Telescope. Answers.com. http://www.answers.com/topic/hubble-space-telescope (accessed December 30, 2008).

Chapter 8

Execute the Design

No plan ever survives first contact with the enemy.

Paraphrased from Helmuth von Moltke

Plans are nothing; planning is everything.

Dwight D. Eisenhower

Introduction

This chapter is a departure from traditional treatments of the design process. Project management (Lewis, 2005) is viewed by many as something separate from the design process. Our view is that the process of organizing and coordinating the effort of a design team is intrinsic to the design process itself. In addition to developing a systems view of the product or service you are developing, we encourage you to develop a systems view of the tasks needed to develop that system. As we shall see, there is a natural linkage between these two views. We have therefore included project management in the Getting Design Right process under the step called "execute the design." Our goal is to deliver on the commitment contained in the mission statement.

The first quotation at the introduction of this chapter suggests that planning is futile because no matter how good the plan looks on paper, events in the real world ("contact with the enemy") are sure to disrupt or nullify the beautiful sequence of steps in the plan. The quote from Eisenhower confirms the outcome ("plans are nothing"), but affirms the essential role of planning. Successful execution requires rapid and intelligent responses to changing circumstances. For responses to be intelligent, they must be based on a systems view of the situation. The best way to develop a systems view is to plan and consider carefully how all of the elements of the plan interact. Thus, "planning is everything."

There are two components to executing the design:

1. Schedule the project and track progress.
2. Conduct management reviews.

A 787 Dream Delayed

When Boeing ran into a 3-week delay on the inaugural launch of its 787 Dreamliner in January 2008, Bank of America analyst Robert Stallard remarked, "Certainly in the decades-long life of the program, [the delay] would prove to be an irrelevance" (Reuters, 2008). Forty-eight weeks later and with still further delays announced, perhaps the relevance increased.

Boeing's 787 Dreamliner (Figure 8.1) is a technological leap into the age of environmental consciousness and fuel economy. Once released, the Dreamliner will be the most fuel-efficient commercial aircraft of its kind, using 20% less fuel and producing 20% fewer emissions than the competition. In addition, engineers have designed a primary structure up to 50% of which is composed of lightweight composite materials. This will be a midsized airplane able to match the flight range of a full-sized jet (Boeing, 2009). The technology behind the Dreamliner may be impressive, but it is the new approach to supply chain management that may have the greatest effect on the outcome of this project.

Rather than following the usual production model wherein parts are ordered from various locations and shipped to Boeing for assembly, the company has devised a new approach. The new production model places a much greater reliance on Boeing's suppliers; whole subsections of the aircraft are being assembled at the vendors' sites before making their way to the Boeing plant in Everett, Washington, for final assembly. The 787 Dreamliner is relying on 135 supply sites to contribute over four million parts from around the world, with subcontractors working in France, Sweden, China, and Japan, to name a few. The project is dependent on each of these orders being accurately assembled and completed on time (Kaste, 2007).

"On time" is the key phrase in this equation. A maiden launch originally scheduled for November 2007 has been postponed at least four times; the prospective launch is now scheduled for late 2009. At the present time, a total of 56 airlines with orders for more than 700 planes at a total price of approximately $144 billion are waiting for the delivery of their products (Boeing, 2009).

Figure 8.1   Boeing 787 Dreamliner aircraft. (Used with permission from the Boeing Company.)

"What's causing the delay?" asks Howard A. Rubel, an analyst at Jeffries and Company. "It's just been a lot of little things" (Wayne and Maynard, 2008). There are many reasons for the delay: a 58-day strike by over 27,000 Boeing employees in the fall of 2008 and a major engineering error that required the replacement of up to 8,000 fasteners, as well as parts shortages and difficulties ramping up the supply chain. "Minding the parts is a big deal," says Bob Noble, vice president of the Boeing 787 supply management. "You've got to be sure that everyone down through the parts stream is ready to move up. ... Sometimes you worry all the way back to the dirt" (Gates, 2008).

The shift in Boeing's supply chain design and assembly process is not inherently the root of such major delays. However, attempting large-scale innovations in both supply systems and product technology may have been an overly ambitious feat for the company to pursue. George T. Haley, director of the Center for International Industry Competitiveness at the University of New Haven concurs: "Boeing has not really mismanaged this effort from a technical perspective, but anytime you have a truly innovative product being developed, especially if it also requires significant innovation in the production process, you are going to run into some potholes that throw your program off schedule" (McInnes, 2008).

Bob Noble, for one, has never lost faith in the airplane or its production system, and there is no doubt that Boeing will submit a truly elegant addition to the skies. For now, however, the skies must wait.

Joanna Jackson

Schedule the Project and Track Progress

There are eight steps in project scheduling:

1. Develop work breakdown structure and task list.
2. Estimate task durations.
3. Track percentages of task completion.
4. Identify task inputs, outputs, and deliverables.
5. Establish task precedence relationships.
6. Determine start and finish times.
7. Display schedule as a Gantt chart.
8. Adjust the start times for resource availability.

What is the First Step in Project Management?

Ed Balys is a former senior vice-president for Concordia Management, a Montreal-based international project management firm specializing in very large construction projects. Ed has served as project manager for projects totaling over $500 million, including the $200 million Museum of Civilization in Ottawa.

At the beginning of every project, Ed's practice is to gather together everyone with a key interest in the project for a 2-day retreat. For a hospital construction project, for example, this would include the senior management of the hospital, but it would also include the doctors in charge of different departments, the head nurse, and the maintenance chief, plus engineers responsible for the facilities and their subsystems.

Ed says that he always begins such a retreat by asking the group what is the most important thing that he, as project manager, needs to do first. Everyone with a special interest in the project will typically suggest something related to his or her primary concern. The engineers might suggest checking the local building codes; the doctor in charge of radiology might suggest placing orders for equipment requiring long lead times, and so on. Ed listens to all their concerns, writing them up on a whiteboard or flipchart for use later in the retreat. Then, after everyone has had a chance to speak, he turns to them with a smile and tells them, "You are all wrong. The first thing I have to do as project manager is to establish the scope of this project."

All projects have a budget. In the case of a hospital project, the senior management may have plans to raise $100 million from donors, but may currently have only $30 million for the project. The scope of the project must be limited to what can be accomplished for $30 million. Not everything that everyone wants (such as a new cardiology wing) can be included in the initial scope of the project. Even though senior management is confident that it can raise the $100 million, the project must fit the available budget. Ed then proceeds to walk the group through a process in which they must decide what is within the scope of the project and what is out of scope. Features that are declared to be out of scope are typically put into a future phase of construction and given a priority. Ed says that the various stakeholders are satisfied as long as they see that their special concerns (the new cardiology wing, for example) have been recognized and assigned some level of priority.

As the fundraising continues and money is raised for additional phases, those features will be constructed according to the schedule of priorities. The work for the current phase is designed in such a way as to make those features possible. Sketches, for example, show where the future cardiology wing will be located. By the end of the 2-day retreat, Ed has created a document detailing the scope of the project and a prioritized schedule of future phases. All the major stakeholders, including senior management, agree to the plan. This critical document becomes the basis for all subsequent decisions on the project (Balys, personal communication 2008).

Develop the Task List (Work Breakdown Structure)

As a design project manager, you define the scope of the project. You must be prepared to identify the work that needs to be done to meet the design goals. If the work exceeds the budget, then you will need to look for economies or to renegotiate with the project owner for more modest goals or for a larger budget. Working on an open-ended project without a clearly defined scope is a prescription for failure and disappointment.

The first step in project scheduling is to develop the list of tasks that need to be accomplished within the project scope. The *work breakdown structure* is a hierarchical list of tasks in which general tasks are progressively broken down into finer levels of detail. Table 8.1 displays a possible work breakdown structure for the toy catapult design project. The higher levels of the structure (design systems; design, build, and test subsystems; integrate subsystems; and verify requirements) form a generic template that is suitable for many design projects. Another, more detailed template for design can be found by reading the table of contents for this text.

From the work breakdown structure, we extract the list of detailed, low-level tasks that must be performed. Table 8.2 displays the list extracted from Table 8.1. Table 8.2 includes a unique

Table 8.1   Work Breakdown Structure for Toy Catapult

Design systems		
	Develop subsystem requirements	
Design, build, and test subsystems		
	Develop projectile launch system	
		Design projectile launch system
		Build projectile launch system
		Test projectile launch system
	Develop energy storage system	
		Design energy storage system
		Build energy storage system
		Test energy storage system
	Develop projectile containment system	
		Design projectile containment system
		Build projectile containment system
		Test projectile containment system
	Develop lock and trigger system	
		Design lock and trigger system
		Build lock and trigger system
		Test lock and trigger system
	Develop body (enclosure system)	
		Design body (enclosure system)
		Build body (enclosure system)
		Test body (enclosure system)
Integrate subsystems into working prototype		
	Integrate launch and storage systems	
	Integrate launch, storage, and containment systems	
	Integrate working system	

(*Continued*)

Table 8.1   Work Breakdown Structure for Toy Catapult (Continued)

Verify system requirements		
	Conduct behavioral tests	
		Test procedure: child plays with toy
		Test procedure: child fails to arm toy
		Test procedure: child releases armed toy near face of self or of another child
		Test procedure: child aims projectile at eyes of self or of another child
		Test procedure: child uses pet rodent as projectile
		Test procedure: the child drops or throws the toy
	Conduct nonbehavioral tests	
		Maximum launch velocity
		Maximum mass launch capability
		Storage volume
		Mean cycles to failure
		Breaking force of critical components
		Repair part cost
		Tensile strength of home-applied adhesive bond
		Time for disassembly and reassembly
		Material cost
		Time to manufacture

activity number for each low-level task. It will be useful in matrix presentations to refer to tasks by their activity numbers rather than by their full names.

Estimate Durations

In order to create a project schedule, it is necessary to estimate the time required, or "duration," of each task. Table 8.3 shows some rough duration estimates for the toy catapult tasks. The estimates are based on prior experience as a hobbyist on similar projects. Experience is usually the best guide. If your design organization tracks actual task times, then you might be able to use these data to estimate future task times better. For example, durations in large software projects are often estimated by multiplying the estimated number of lines of code that need to be written by a standard amount of programming time required per line of code.

The total of these durations is an estimate of the total work required to complete the project. We have estimated that the total project would require 18.55 hours, 11 hours of which involve

Table 8.2 A Task List for Toy Catapult Design Project

Activity No.	Activity Name
A.01	Develop subsystem requirements
A.02	Design projectile launch system
A.03	Build projectile launch system
A.04	Test projectile launch system
A.05	Design energy storage system
A.06	Build energy storage system
A.07	Test energy storage system
A.08	Design projectile containment system
A.09	Build projectile containment system
A.10	Test projectile containment system
A.11	Design lock and trigger system
A.12	Build lock and trigger system
A.13	Test lock and trigger system
A.14	Design body (enclosure system)
A.15	Build body (enclosure system)
A.16	Test body (enclosure system)
A.17	Integrate launch and storage systems
A.18	Integrate launch, storage, and containment systems
A.19	Integrate working system
TP.01	Test procedure: child plays with toy
TP.02	Test procedure: child fails to arm toy
TP.03	Test procedure: child releases armed toy near face of self or of another child
TP.04	Test procedure: child aims projectile at eyes of self or of another child
TP.05	Test procedure: child uses pet rodent as projectile
TP.06	Test procedure: the child drops or throws the toy
TP.07	Maximum launch velocity
TP.08	Maximum mass launch capability

(Continued)

Table 8.2   A Task List for Toy Catapult Design Project (Continued)

Activity No.	Activity Name
TP.09	Storage volume
TP.10	Mean cycles to failure
TP.11	Breaking force of critical components
TP.12	Repair part cost
TP.13	Tensile strength of home-applied adhesive bond
TP.14	Time for disassembly and reassembly
TP.15	Material cost
TP.16	Time to manufacture

testing to make sure the toy satisfies the requirements. "Getting design right" implies a substantial commitment to testing.

Track Percent Complete

As design project manager, you should track the progress of the various tasks to ensure that the project stays on schedule. One way of doing this is to estimate the percent completion of each task (or require the different team leaders to report percent completion to you). Figure 8.2 displays a table designed to keep track of these individual task percentage completions. The spreadsheet on which this table is based includes spinner buttons (in the column labeled "change"). Each spinner

Table 8.3   Estimate Task Durations

Activity No.	Activity Name	Duration (Hours)
A.01	Develop subsystem requirements	1
A.02	Design projectile launch system	0.5
A.03	Build projectile launch system	0.5
A.04	Test projectile launch system	0.1
A.05	Design energy storage system	0.5
A.06	Build energy storage system	0.5
A.07	Test energy storage system	0.1
A.08	Design projectile containment system	0.5
A.09	Build projectile containment system	0.4

Table 8.3   Estimate Task Durations (Continued)

Activity No.	Activity Name	Duration (Hours)
A.10	Test projectile containment system	0.1
A.11	Design lock and trigger system	0.5
A.12	Build lock and trigger system	0.75
A.13	Test lock and trigger system	0.3
A.14	Design body (enclosure system)	0.4
A.15	Build body (enclosure system)	0.5
A.16	Test body (enclosure system)	0.1
A.17	Integrate launch and storage systems	0.2
A.18	Integrate launch, storage, and containment systems	0.2
A.19	Integrate working system	0.4
TP.01	Test procedure: child plays with toy	0.5
TP.02	Test procedure: child fails to arm toy	0.2
TP.03	Test procedure: child releases armed toy near face of self or of another child	0.4
TP.04	Test procedure: child aims projectile at eyes of self or of another child	0.4
TP.05	Test procedure: child uses pet rodent as projectile	0.4
TP.06	Test procedure: the child drops or throws the toy	0.4
TP.07	Maximum launch velocity	1
TP.08	Maximum mass launch capability	0.2
TP.09	Storage volume	0.1
TP.10	Mean cycles to failure	5
TP.11	Breaking force of critical components	1
TP.12	Repair part cost	0.1
TP.13	Tensile strength of home-applied adhesive bond	1
TP.14	Time for disassembly and reassembly	0.1
TP.15	Material cost	0.1
TP.16	Time to manufacture	0.1
	Total	18.55

Activity Number	Activity Name	Duration (Hrs)	Percent Complete (0–100)	Change Percent Complete	Work Completed	Work Remaining
A.01	Develop Subsystem Requirements	1	50	◀ ▶	0.5	0.5
A.02	Design Projectile Launch System	0.5	0	◀ ▶	0	0.5
A.03	Build Projectile Launch System	0.5	0	◀ ▶	0	0.5
A.04	Test Projectile Launch System	0.1	0	◀ ▶	0	0.1
A.05	Design Energy Storage System	0.5	0	◀ ▶	0	0.5
A.06	Build Energy Storage System	0.5	0	◀ ▶	0	0.5
A.07	Test Energy Storage System	0.1	0	◀ ▶	0	0.1
A.08	Design Projectile Containment System	0.5	0	◀ ▶	0	0.5
A.09	Build Projectile Containment System	0.4	0	◀ ▶	0	0.4
A.10	Test Projectile Containment System	0.1	0	◀ ▶	0	0.1
A.11	Design Lock and Trigger System	0.5	0	◀ ▶	0	0.5
A.12	Build Lock and Trigger System	0.75	0	◀ ▶	0	0.75
A.13	Test Lock and Trigger System	0.3	0	◀ ▶	0	0.3
A.14	Design Body (Enclosure System)	0.4	0	◀ ▶	0	0.4
A.15	Build Body (Enclosure System)	0.5	0	◀ ▶	0	0.5
A.16	Test Body (Enclosure System)	0.1	0	◀ ▶	0	0.1
A.17	Integrate Launch and Storage Systems	0.2	0	◀ ▶	0	0.2
A.18	Integrate Launch, Storage, and Containment Systems	0.2	0	◀ ▶	0	0.2
A.19	Integrate Working System	0.4	0	◀ ▶	0	0.4
TP.01	Test Procedure: Child plays with toy	0.5	0	◀ ▶	0	0.5
TP.02	Test Procedure: Child fails to arm toy.	0.2	0	◀ ▶	0	0.2
TP.03	Test Procedure: Child releases armed toy near face of self or another child	0.4	0	◀ ▶	0	0.4
TP.04	Test Procedure: Child aims projectile at eyes of self or of another child	0.4	0	◀ ▶	0	0.4
TP.05	Test Procedure: Child uses pet rodent as projectile	0.4	0	◀ ▶	0	0.4
TP.06	Test Procedure: The child drops or throws the toy.	0.4	0	◀ ▶	0	0.4
TP.07	Maximum launch velocity	1	0	◀ ▶	0	1
TP.08	Maximum mass launch capability	0.2	0	◀ ▶	0	0.2
TP.09	Storage volume	0.1	0	◀ ▶	0	0.1
TP.10	Mean cycles to failure	5	0	◀ ▶	0	5
TP.11	Breaking force of critical components	1	0	◀ ▶	0	1
TP.12	Repair part cost	0.1	0	◀ ▶	0	0.1
TP.13	Tensile strength of home-applied adhesive bond	1	0	◀ ▶	0	1
TP.14	Time for disassembly and reassembly	0.1	0	◀ ▶	0	0.1
TP.15	Material cost	0.1	0	◀ ▶	0	0.1
TP.16	Time to manufacture	0.1	0	◀ ▶	0	0.1
	Total	18.55			0.5	18.05

Figure 8.2 Track percentage completion of each task.

is linked to one of the percent completion cells. As you click on the spinner, right or left, the corresponding percent completion increases or decreases, respectively. This makes it easy to update the progress of the project. The additional columns in Figure 8.2 compute the work completed and the work remaining based on the duration and percent completed columns. Cells containing formulas in this spreadsheet are shaded.

Identify Task Inputs, Outputs, and Deliverables

Dependencies among tasks are important to keep in mind. Even the simple sequence of design, build, and test implies that you cannot build what you have not designed and you cannot fully test something that is not built. It is useful to construct some sort of a map as to what is needed from one task for another task to begin. A systematic way to study this is to create a *task input–output matrix:* a square matrix of the tasks in which the cells of the matrix describe what output from one task (the row task) is needed as input for another task (the column task) to begin.

Tables 8.4–8.6 are snapshots of different regions of a task input–output matrix for the toy catapult. The matrix is too large to be readable if reproduced on a single page. We identified that the various design activities (A.02, A.05, A.08, and A.11) should not proceed until they receive a list of requirements for their subsystem and specifications of the interfaces with other subsystems. Consequently, an important document will be the "subsystem requirements and interface specifications document." This is the major output of task A.01, "develop subsystem requirements."

Table 8.4 Task Inputs and Outputs

(Row) Task Supplies to (Column)Task	Develop Subsystem Requirements A.01	Design Projectile Launch System A.02	Build Projectile Launch System A.03	Test Projectile Launch System A.04	Design Energy Storage System A.05
Develop subsystem requirements A.01		subsystem requirements and interface specifications			subsystem requirements and interface specifications
Design projectile launch system A.02			drawings		
Build projectile launch system A.03				subassembly	
Test projectile launch system A.04					
Design energy storage system A.05					

Table 8.5 Task Inputs and Outputs

(Row) Task Supplies to (Column) Task	Storage Volume	Mean Cycles to Failure	Breaking Force of Critical Components	Repair Part Cost	Tensile Strength of Home-Applied Adhesive Bond	Time for Disassembly and Reassembly
	TP.09	TP.10	TP.11	TP.12	TP.13	TP.14
Develop subsystem requirements A.01						
Design projectile launch system A.02	preliminary dimensions			bill of materials		process plan
Build projectile launch system A.03						
Test projectile launch system A.04			approved bill of materials		approved bill of materials	
Design energy storage system A.05	preliminary dimensions			bill of materials		process plan

Table 8.6   Task Inputs and Outputs

(Row) Task Supplies to (Column) Task		*Integrate Launch and Storage Systems*	*Integrate Launch, Storage, and Containment Systems*	*Integrate Working System*	*Test Procedure: Child Plays with Toy*
		A.17	**A.18**	**A.19**	**TP.01**
Integrate launch and storage systems	A.17		tested subassembly		
Integrate launch, storage, and containment systems	A.18			tested subassembly	
Integrate working system	A.19				working prototype
Test procedure: child plays with toy	TP.01				
Test procedure: child fails to arm toy	TP.02				approval of safety
Test procedure: child releases armed toy near face of self or of another child	TP.03				approval of safety
Test procedure: child aims projectile at eyes of self or of another child	TP.04				approval of safety

The main outputs of the design activities are drawings and preliminary dimensions. The drawings are used as inputs to the various build tasks. The preliminary dimensions are used for designing the enclosure and for conducting the storage volume test. We also discover that a number of the test activities (TP.12 and TP.15) will need bills of material. Other test activities (TP.14 and TP.16) will need process plans (descriptions of how the subsystems are to be constructed). Logically, these are outputs of the design tasks. Two of the test activities (TP.11 and TP.13) are physical tests of components. We decided to wait for the tests of the relevant subassemblies to complete before starting these test activities. In a sense, then, we are requiring an "approved bill of materials" (approved by passing the initial tests) as input to these tests.

The various build tasks create physical subassemblies that are the inputs to test activities. The test activities output "tested subassemblies." These tested subassemblies are integrated into working partial prototypes. Several of the tests (TP.02, TP.03, TP.04, TP.07, and TP.08) require only a working partial prototype as input. The remaining tests (TP.01, TP.05, TP.06, and TP.10) require a full working prototype, the output of A.19. We also decided to delay the behavior test involving an actual child (TP.01) until after the safety tests (TP.02, TP.03, and

Table 8.7 Summarize Deliverables and Input–Output Relationships

Deliverable Type	Required as Output from Task(s)	Required as Input to Task(s)
Subsystem requirements and interface specifications	A.01	A.02, A.05, A.08, A.11, A.14
Drawings	A.02, A.05, A.08, A.11, A.14	A.03, A.06, A.09, A.12
Subassembly	A.03, A.06, A.09, A.12	A.07, A.10, A.13
Preliminary dimensions	A.02, A.05, A.08, A.11	A.14, TP.09
Tested subassembly	A.04, A.07, A.10, A.13, A.17, A.18	A.17, A.18, A.19
Finalized dimensions	A.18	A.15
Bill of materials	A.02, A.05, A.08, A.11, A.14	TP.12, TP.15
Approved bill of materials	A.04	TP.11, TP.13
Process plan	A.02, A.05, A.08, A.11, A.14	TP.14, TP.16
Working partial prototype	A.18	TP.02, TP.03, TP.04, TP.07, TP.08
Prototype	A.15	A.16
Approval of safety	TP.02, TP.03, TP.04	TP.01
Working prototype	A.19	TP.01, TP.05, TP.06

TP.04) have been conducted. In a sense, then, TP.01 requires as input a set of safety approvals output from these safety tests.

From the task input–output matrix, we extract a list of deliverables and summarize the input–output relationships as shown in Table 8.7 for the toy catapult. This is a systems view of the design process. A *deliverable* is something (tangible or intangible) produced as a result of a task or project. It could be a design drawing, a physical prototype, or a test result. Creating the task input–output matrix has given us a systems view. Any particular team working on the project needs to know only the inputs and outputs for the particular task assigned to it. As design project manager, it is your responsibility to ensure that each team knows the complete set of deliverables for which it is responsible. For example, we discovered the need for process plans in order to conduct tests TP.14 and TP.16. We have included responsibility for creating these process plans as part of the appropriate design tasks. These process plans are deliverables of the design teams.

Establish Task Precedence Relationships

From the task input–output matrix, it is straightforward to create a *task precedence matrix*—a square matrix using the list of tasks as row and column headings, in which a cell entry of "1" indicates that the row task must precede the column task. In general, if the task input–output

Table 8.8 Precedence Relationships before Reordering

	A.01	A.02	A.03	A.04	A.05	A.06	A.07	A.08	A.09	A.10	A.11	A.12	A.13	A.14	A.15	A.16	A.17	A.18	A.19	TP.01	TP.02	TP.03	TP.04	TP.05	TP.06	TP.07	TP.08	TP.09	TP.10	TP.11	TP.12	TP.13	TP.14	TP.15	TP.16
A.01		1			1			1			1			1																					
A.02			1											1														1			1		1	1	1
A.03				1																															
A.04																	1													1		1			
A.05						1								1														1			1		1	1	1
A.06							1										1																		
A.07																																			
A.08									1					1														1			1		1	1	1
A.09										1																									
A.10																		1																	
A.11												1		1														1			1		1	1	1
A.12													1																						
A.13																			1																
A.14															1													1			1		1	1	1
A.15																1																			
A.16																																			
A.17																		1																	

(Continued)

Table 8.8 Precedence Relationships before Reordering (Continued)

	A.01	A.02	A.03	A.04	A.05	A.06	A.07	A.08	A.09	A.10	A.11	A.12	A.13	A.14	A.15	A.16	A.17	A.18	A.19	TP.01	TP.02	TP.03	TP.04	TP.05	TP.06	TP.07	TP.08	TP.09	TP.10	TP.11	TP.12	TP.13	TP.14	TP.15	TP.16
A.18															1				1		1	1	1			1	1								
A.19																				1				1	1				1						
TP.01																																			
TP.02																				1															
TP.03																				1															
TP.04																				1															
TP.05																																			
TP.06																																			
TP.07															1																				
TP.08															1																				
TP.09																																			
TP.10																																			
TP.11																																			
TP.12																																			
TP.13																																			
TP.14																																			
TP.15																																			
TP.16																																			

matrix indicates a deliverable in a cell, there would be a "1" in the corresponding cell of the task precedence matrix. Table 8.8 is a task precedence matrix based on the task input–output matrices of Tables 8.4–8.6. The task precedence matrix, with only "1s" and blanks as entries, is more convenient for making schedule calculations than the wordier task input–output matrix.

Observe that we have shaded cells along the main diagonal of Table 8.8. Entries below the diagonal of the matrix indicate that a task cannot start until some task farther down on this list finishes. For example, we have decided that building the body (A.15) should wait until there is a working partial prototype (the output of A.18) and the basic performance requirements for launch velocity (TP.07) and launch capacity (TP.08) have been verified. These decisions result in "1s" below the main diagonal in the A.15 column. This raises a question: "Is there a sequence in which all of the tasks can be performed without violating any precedence relationship?" Put another way, the question is whether the order of the rows and columns of the task precedence matrix can be changed so that all of the "1s" are in cells above the main diagonal. The answer to both questions may be "no" if there is a cycle of precedence constraints (for example, task A precedes task B, which precedes task C, which precedes task A).

In the spreadsheet that accompanies this chapter, we have included a macro written in Visual Basic that attempts to "upper-triangularize" a selected matrix. That is, it attempts to reorganize the rows and columns of a selected matrix so that all of the nonblank entries in the matrix lie above the main diagonal.

The macro will create a new ordering to the rows and columns. If there are no cycles, then all of the nonblank entries will lie above the main diagonal. Table 8.9 shows the results of applying the macro to the matrix from Table 8.8. Note that the descriptive headings in row 1 and column A are updated automatically because these cells contain formulas that look up the descriptive task names based on the task numbers in the row and column headings for the matrix.

Table 8.10 is a clearer view of the revised sequence of tasks from Table 8.9. The list of tasks in Table 8.10 has the property that if the tasks are performed in this sequence, no precedence relationship will be violated. The original list of tasks (Table 8.2) did not have this property.

If the matrix cannot be reordered into upper triangular form, then a cycle of precedence constraints is infeasible. As design project manager, you should revise the task list to eliminate such cycles. For example, if two design activities cannot start until the other one completes, this means that an interface between the two subsystems is not completely defined. Introduce a new task to define the interface so that the design activities can proceed independently of each other.

Schedule the Project

Once the task precedence matrix has been reorganized into upper triangular form, we can use spreadsheet formulas to compute a *schedule,* which is simply a listing of start and finish times for each task in the task list. An *earliest start schedule* is a schedule in which, beginning from a project start time, every task starts as early as possible, subject only to task durations and the task precedence matrix.

Table 8.9 Precedence Matrix in Upper Triangular Form

		C	D	E	F	G	H	I	J	K	L	M	N	O	P	Q	R	S	T	U	V	W	X	Y	Z	AA	AB	AC	AD	AE	AF	AG	AH	AI	AJ	AK
B		A.01	A.02	A.03	A.04	A.05	A.06	A.07	A.08	A.09	A.10	A.11	A.12	A.13	A.14	A.17	A.18	TP.07	TP.08	A.15	A.16	A.19	TP.02	TP.03	TP.04	TP.01	TP.05	TP.06	TP.09	TP.10	TP.11	TP.12	TP.13	TP.14	TP.15	TP.16
3	A.01		1			1			1			1			1																					
4	A.02			1											1																		1		1	1	1
5	A.03				1																													1			
6	A.04															1																1					
7	A.05						1								1															1			1		1	1	1
8	A.06							1																													
9	A.07															1																					
10	A.08									1					1															1			1		1	1	1
11	A.09										1																										
12	A.10																	1																			
13	A.11												1		1															1			1		1	1	1
14	A.12														1								1														
15	A.13																				1																
16	A.14																													1			1		1	1	1
17	A.17																	1			1		1														
18	A.18																				1			1	1	1											
19	TP.07																		1	1	1																

Table 8.9 Precedence Matrix in Upper Triangular Form (Continued)

B	C	D	E	F	G	H	I	J	K	L	M	N	O	P	Q	R	S	T	U	V	W	X	Y	Z	AA	AB	AC	AD	AE	AF	AG	AH	AI	AJ	AK
	A.01	A.02	A.03	A.04	A.05	A.06	A.07	A.08	A.09	A.10	A.11	A.12	A.13	A.14	A.17	A.18	TP.07	TP.08	A.15	A.16	A.19	TP.02	TP.03	TP.04	TP.01	TP.05	TP.06	TP.09	TP.10	TP.11	TP.12	TP.13	TP.14	TP.15	TP.16
20 TP.08																			1																
21 A.15																				1															
22 A.16																																			
23 A.19																									1	1	1		1						
24 TP.02																									1		1								
25 TP.03																									1										
26 TP.04																									1										
27 TP.01																																			
28 TP.05																																			
29 TP.06																																			
30 TP.09																																			
31 TP.10																																			
32 TP.11																																			
33 TP.12																																			
34 TP.13																																			
35 TP.14																																			
36 TP.15																																			
37 TP.16																																			

Table 8.10   Revised Sequence of Tasks

Sequence	Task	
1	Develop subsystem requirements	A.01
2	Design projectile launch system	A.02
3	Build projectile launch system	A.03
4	Test projectile launch system	A.04
5	Design energy storage system	A.05
6	Build energy storage system	A.06
7	Test energy storage system	A.07
8	Design projectile containment system	A.08
9	Build projectile containment system	A.09
10	Test projectile containment system	A.10
11	Design lock and trigger system	A.11
12	Build lock and trigger system	A.12
13	Test lock and trigger system	A.13
14	Design body (enclosure system)	A.14
15	Integrate launch and storage systems	A.17
16	Integrate launch, storage, and containment systems	A.18
17	Maximum launch velocity	TP.07
18	Maximum mass launch capability	TP.08
19	Build body (enclosure system)	A.15
20	Test body (enclosure system)	A.16
21	Integrate working system	A.19
22	Test procedure: child fails to arm toy	TP.02
23	Test procedure: child releases armed toy near face of self or of another child	TP.03
24	Test procedure: child aims projectile at eyes of self or of another child	TP.04
25	Test procedure: child plays with toy	TP.01
26	Test procedure: child uses pet rodent as projectile	TP.05
27	Test procedure: child drops or throws the toy	TP.06

Table 8.10   Revised Sequence of Tasks (Continued)

Sequence	Task	
28	Storage volume	TP.09
29	Mean cycles to failure	TP.10
30	Breaking force of critical components	TP.11
31	Repair part cost	TP.12
32	Tensile strength of home-applied adhesive bond	TP.13
33	Time for disassembly and reassembly	TP.14
34	Material cost	TP.15
35	Time to manufacture	TP.16

Compute an Earliest Start Schedule (Advanced Topic)

This section may be skipped without affecting the flow of the chapter.

Three types of formulas are needed to compute an earliest start schedule. We refer to tasks by their sequence number: an index i or j running from 1 up to the number of tasks in the order given by the upper triangular form of the precedence matrix. The first formula is that the finish time of a task with index j is its start time plus its duration:

$$\mathrm{FinishTime}(j) = \mathrm{StartTime}(j) + \mathrm{Duration}(j).$$

The second type of formula is for the delay in the start time of task j caused by the need to wait for the finish of task i. We refer to this as the *precedence delay* of task i on task j. Because the precedence matrix has a 1 in cell (i,j), if task j must wait for the finish of task i, we can simply multiply this cell of the precedence matrix by the finish time of task i to compute the precedence delay:

$$\mathrm{PrecedenceDelay}(i,j) = \mathrm{PrecedenceMatrix}(i,j) * \mathrm{FinishTime}(i).$$

Because the task precedence matrix is in upper triangular form, we do not need to consider precedence delays on task j from any task index greater than j. (That is, $\mathrm{PrecedenceMatrix}(i,j) = 0$ if $i > j$.) For task j, we must consider all the precedence delays for tasks with index i less than j. That is, we must consider the list $\mathrm{PrecedenceDelay}(1,j)$, $\mathrm{PrecedenceDelay}(2,j)$, $\mathrm{PrecedenceDelay}(3,j)$, and so on, up to $\mathrm{PrecedenceDelay}(j-1,j)$. A shorthand notation for this list is

$$\mathrm{PrecedenceDelay}(1,j): \mathrm{PrecedenceDelay}(j-1,j).$$

The symbol ":" means "up to" in this context. For the first task, $j = 1$, there can be no precedence delay, so we interpret this to be an empty list.

The third type of formula expresses the fact that the earliest start time of an activity with index j is simply the project start time plus the maximum of all precedence delays from tasks with lower index numbers:

$$\mathrm{StartTime}(j) = \mathrm{Max}(\mathrm{PrecedenceDelay}(1,j): \mathrm{PrecedenceDelay}(j-1,j).$$

For the first task, the maximum of an empty list is zero, so the start time of the first task will exactly equal the project start time. Its finish time will be determined by the first formula, and this finish time will cause precedence delays for all later tasks in the column for which there is a 1 in the first row of the precedence matrix.

All of these formulas can be implemented in a spreadsheet, but the details are outside the scope of this text to explain. The spreadsheet file that accompanies this chapter demonstrates the formulas required. All shaded cells in the spreadsheet contain formulas relevant to computing and displaying schedules. Figure 8.3 shows the spreadsheet used to compute an earliest start schedule for the toy catapult example. Cells in spreadsheet column "E" contain formulas of the first type, for computing finish times. Cells in the upper triangular portion of the range from "H3" to "AP37" contain formulas of the second type, for computing precedence delays. These formulas make reference to the task precedence matrix cells on a separate worksheet. Finally, cells in spreadsheet column "B" use formulas of the third type, for computing start times.

As an example of the calculations, consider task A.19, "integrate working system." Referring to the task precedence matrix in Table 8.9, we see that task A.19 corresponds to column "W" of that worksheet. There are two "1s" in that column, indicating that A.19 must wait until A.13 and A.18 finish. Referring to column "E" of Figure 8.3, we see that these finish times are 2.55 and 2.5, respectively. These are precisely the numbers that appear as positive precedence delays in column "AB" of Figure 8.3, although the 2.55 has been rounded to 2.6 in the display. The rest of the precedence delays in column "AB" are zero. The maximum precedence delay is therefore 2.55 (the maximum of cells in the range "AB3:AB22"). Because the project start time is zero, the earliest start time of A.19 should be 2.55. That is precisely the number computed in cell "B23," the start time of A.19.

Columns "B" and "E" contain the start and finish times of the tasks corresponding to the earliest start schedule for the toy catapult project. Rather than discuss this schedule here, we first display it as a Gantt chart.

Display a Gantt Chart

For small projects (30 activities) or for the major activities of a large project, it is useful to display schedules in the form of a Gantt chart. Figure 8.4 is a *Gantt chart* of the tasks for the toy catapult project. Each task occupies a separate row of the chart. The horizontal axis measures time in hours from the start of the project. Each task is displayed as a horizontal bar from its start time on the horizontal axis to its finish time. The bar is divided into two shades to show the fraction of work completed. The Gantt chart was first published by Henry L. Gantt in 1910.

Gantt charts are not one of the chart types supported by MS Excel. However, a simple trick can be employed to use the stacked bar chart in MS Excel to display Gantt charts (spreadsheet skill: display Gantt chart).

Now that the schedule is displayed in a Gantt chart form, we can discuss more easily the type of information communicated in an earliest start schedule. Recall from adding up the durations in Table 8.3 that the toy catapult project would take 18.55 hours to complete if we performed the tasks serially, one after another. Note that if we performed the operations serially, we would have to follow the order specified in Table 8.10 to ensure that all the precedence relationships are satisfied. The earliest start schedule reveals the minimum possible time to complete the project

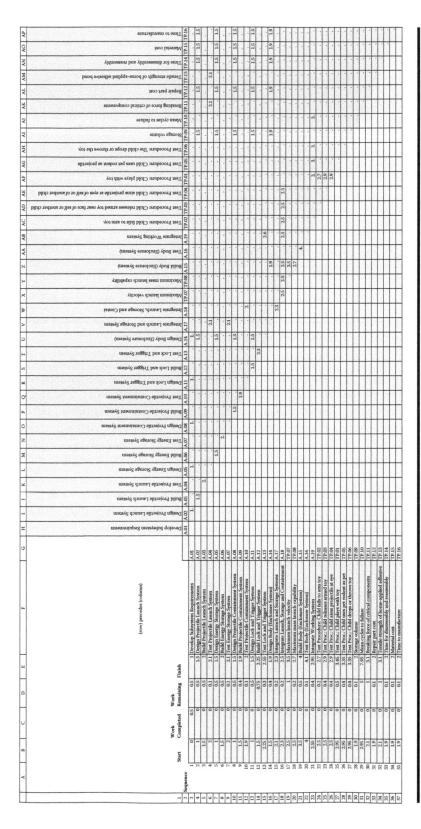

Figure 8.3 Determine start and finish times.

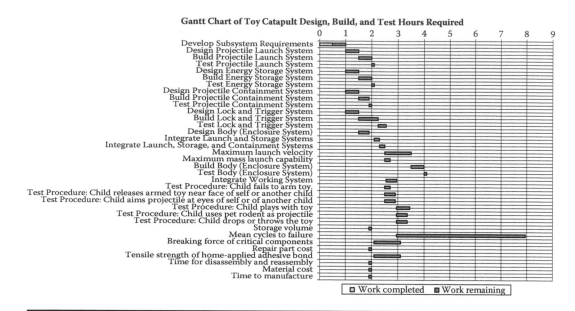

Figure 8.4 Gantt chart of toy catapult design, build, and test.

and satisfy all of the precedence restrictions. As the Gantt chart reveals, the earliest start schedule for the toy catapult finishes when the mean cycles to failure test is complete (TP.09). This occurs 7.95 hours after the project start. Rather than assuming that tasks must be performed serially, the earliest start schedule assumes tasks can be performed in parallel (simultaneously) unless there is a precedence relationship between the tasks. You can see the power of *working in parallel*. The project time can be reduced by more than 50% (from 18.55 to 7.95 hours) if it is possible to conduct some tasks in parallel.

In general, the earliest start schedule is unrealistic. Often, it is not possible to work in parallel to the degree exploited by the earliest start schedule. For example, observe from the Gantt chart in Figure 8.4 that many of the test procedures are scheduled to occur simultaneously. Test procedures TP.01 "child plays with toy," TP.05 "child uses pet rodent," and TP.06 "child drops or throws toy" are all supposed to happen in parallel. Clearly, because we plan to build only a single prototype and the prototype can be used in only one test at a time, this schedule is infeasible. The problem is that earliest start schedules ignore *resource constraints* such as team availability, testing equipment availability, and prototype availability.

The challenge is to find a schedule that satisfies precedence relationships, as well as resource constraints. This is called *resource-constrained scheduling*.

Adjust the Schedule for Team Availability

As suggested in the previous section, it may be necessary to modify an earliest start schedule to account for limited resources. Our approach will be to add delays to start times and to use a trial and error method to adjust these delays to satisfy resource constraints. That is sufficient for small projects. For large projects, you will want to use project scheduling software to search for optimized solutions.

	A	B	C	D	E	F	G	H
1	**Table used for Gantt Chart**	Sequence	Change Delay		Start	Work Completed	Work Remaining	Finish
2								
3	Develop Subsystem Requirements				0	0.5	0.5	1
4	Design Projectile Launch System	2	◀ ▶	0	1	0	0.5	1.5
5	Build Projectile Launch System	3	◀ ▶	2	2	0	0.5	2.5
6	Test Projectile Launch System	4	◀ ▶	0	2.5	0	0.1	2.6
7	Design Energy Storage System	5	◀ ▶	0	1	0	0.5	1.5
8	Build Energy Storage System	6	◀ ▶	0	1.5	0	0.5	2
9	Test Energy Storage System	7	◀ ▶	0	2	0	0.1	2.1
10	Design Projectile Containment System	8	◀ ▶	2	1.5	0	0.5	2
11	Build Projectile Containment System	9	◀ ▶	2	2.5	0	0.4	2.9
12	Test Projectile Containment System	10	◀ ▶	0	2.9	0	0.1	3
13	Design Lock and Trigger System	11	◀ ▶	0	1	0	0.5	1.5
14	Build Lock and Trigger System	12	◀ ▶	0	1.5	0	0.75	2.25
15	Test Lock and Trigger System	13	◀ ▶	0	2.25	0	0.3	2.55
16	Design Body (Enclosure System)	14	◀ ▶	0	2	0	0.4	2.4
17	Integrate Launch and Storage Systems	15	◀ ▶	0	2.6	0	0.2	2.8
18	Integrate Launch, Storage, and Containment System	16	◀ ▶	0	3	0	0.2	3.2
19	Maximum launch velocity	17	◀ ▶	13	6.45	0	1	7.45
20	Maximum mass launch capability	18	◀ ▶	0	3.2	0	0.2	3.4
21	Build Body (Enclosure System)	19	◀ ▶	6	8.95	0	0.5	9.45
22	Test Body (Enclosure System)	20	◀ ▶	0	9.45	0	0.1	9.55
23	Integrate Working System	21	◀ ▶	0	3.2	0	0.4	3.6
24	Test Procedure: Child fails to arm toy	22	◀ ▶	0	3.2	0	0.2	3.4
25	Test Procedure: Child releases armed toy near face of self or another child	23	◀ ▶	0	3.2	0	0.4	3.6
26	Test Procedure: Child aims projectile at eyes of self or of another child	24	◀ ▶	2	3.7	0	0.4	4.1
27	Test Procedure: Child plays with toy	25	◀ ▶	4	5.1	0	0.5	5.6
28	Test Procedure: Child uses pet rodent as projectile	2	◀ ▶	2	4.1	0	0.4	4.5
29	Test Procedure: Child drops or throws the toy	27	◀ ▶	6	5.1	0	0.4	5.5
30	Storage volume	28	◀ ▶	0	2.4	0	0.1	2.5
31	Mean cycles to failure	29	◀ ▶	0	3.6	0	5	8.6
32	Breaking force of critical components	30	◀ ▶	0	2.6	0	1	3.6
33	Repair part cost	31	◀ ▶	0	2.4	0	0.1	2.5
34	Tensile strength of home-applied adhesive bond	32	◀ ▶	0	2.6	0	1	3.6
35	Time for disassembly and reassembly	33	◀ ▶	0	2.4	0	0.1	2.5
36	Material cost	34	◀ ▶	0	2.4	0	0.1	2.5
37	Time to manufacture	35	◀ ▶	0	2.4	0	0.1	2.5
38								

Figure 8.5   Inserting start time delays.

Figure 8.5 shows how we have modified the spreadsheet for computing earliest start schedules to include a column for delays, measured in quarter-hours. The formula for start times is modified to add in the delay (after converting to hours):

$$\text{StartTime}(j) = \text{Delay}(j)/4 + \text{Max}(\text{PrecedenceDelay}\ (1,j): \text{PrecedenceDelay}\ (j-1,j).$$

This is the formula used in column "E" of the spreadsheet of Figure 8.5. The resulting schedule will satisfy the precedence constraints, but the insertion of delays permits us to satisfy other constraints as well. If any of the delays are nonzero, then the schedule ceases to be an earliest start time schedule.

Observe in Figure 8.5 that we have added spinner buttons to the spreadsheet. Each of these spinner buttons is linked to one of the delay cells. In this way, we can simply click the spinner, right or left, to increase or decrease the delay in starting a particular task. The rest of the schedule adjusts automatically, and the Gantt chart changes dynamically.

With this capability to stagger the start times of the tasks easily, we now consider ways to make the schedule more realistic. For this text, we consider only a single type of resource: the availability of a team to perform the work. Suppose, for the toy catapult project, we have five teams available with basic responsibilities as shown in Table 8.11. To keep matters simple, we assume there is only a single person in each team (the team leader).

Table 8.11   Teams Available for Toy Catapult Project

Team No.	Team Leader	Description
1	Marie	Designer for catapult
2	Xin	Builder for catapult
3	Amar	Designer for lock and trigger system and body
4	Yolanda	Builder for lock and trigger system
5	Frederic	Testing and final build

The next step is to assign each task to exactly one team. That team is responsible for completing the task according to the schedule. This is done using a worksheet such as that shown in Figure 8.6. The shaded cells contain various table lookup formulas. The white cells are where we enter the team number of the team responsible for each task. The team leader name shows up automatically in the next column, and a name for the responsibility is created in the next column by concatenating the name of the team leader with the description of the task.

Observe that we have assigned all the test activities that require the prototype to a single team. In that way, if we find a schedule that satisfies team availability it will also satisfy prototype availability.

The next step is to create a Gantt chart in which the responsibilities are sorted by task sequence within team number. It is outside the scope of this text to describe the spreadsheet formulas to do this. The spreadsheet that accompanies this chapter is a sample implementation. Figure 8.7 shows the resulting Gantt chart when all the delays are set to zero. Observe that all of Marie's responsibilities are listed first, then Xin's responsibilities, and so on to Frederic's responsibilities. By grouping the tasks in this way, it is easier to see whether tasks for the same person overlap in time.

Now we must use trial and error to stagger the start times (add delays) to make sure that no team is expected to be working on more than one task at a time, but that the project is still completed as early as is feasible. This is now a computer puzzle game: Click the spinner buttons and study the resulting responsibility Gantt chart.

We begin with team 1, Marie, and note that she cannot be expected to design the projectile launch system, the energy storage system, and the projectile containment system simultaneously in the time it takes to design just one of them. But this is what the schedule in Figure 8.7 has her doing.

Figure 8.8 is the result of adjusting start times to ensure that each team works on only one task at a time. Table 8.12 is the corresponding schedule showing the inserted delay times. Observe that we have spread Marie's work out so that it should now be feasible for her to accomplish the tasks. You can visually check each team's schedule to see that it is feasible, that is, that it has no overlapping tasks.

The time to complete the project is now estimated to be 13.6 hours. This is much longer than the 7.95 hours of the earliest start schedule, but it is a more realistic schedule because it considers team availability. It is shorter than the purely serial time of 18.5 hours because we have exploited the concept of parallel work by having different teams work on the different subsystems. However, because we have only a single prototype, it is not possible to conduct many of the tests in parallel.

Figure 8.8 also reveals another concern with managing a design project. Working in parallel can shorten the overall design project time, but it can be difficult to keep everyone busy.

Sequence	Task	Task Name	Assigned to Team Number	Team Leader	Responsibilities
1	A.01	Develop Subsystem Requirements	5	Frederic	Frederic: Develop Subsystem Requirements
2	A.02	Design Projectile Launch System	1	Marie	Marie: Design Projectile Launch System
3	A.03	Build Projectile Launch System	2	Xin	Xin: Build Projectile Launch System
4	A.04	Test Projectile Launch System	2	Xin	Xin: Test Projectile Launch System
5	A.05	Design Energy Storage System	1	Marie	Marie: Design Energy Storage System
6	A.06	Build Energy Storage System	2	Xin	Xin: Build Energy Storage System
7	A.07	Test Energy Storage System	2	Xin	Xin: Test Energy Storage System
8	A.08	Design Projectile Containment System	1	Marie	Marie: Design Projectile Containment System
9	A.09	Build Projectile Containment System	2	Xin	Xin: Build Projectile Containment System
10	A.10	Test Projectile Containment System	2	Xin	Xin: Test Projectile Containment System
11	A.11	Design Lock and Trigger System	3	Amar	Amar: Design Lock and Trigger System
12	A.12	Build Lock and Trigger System	4	Yolanda	Yolanda: Build Lock and Trigger System
13	A.13	Test Lock and Trigger System	4	Yolanda	Yolanda: Test Lock and Trigger System
14	A.14	Design Body (Enclosure System)	2	Amar	Amar: Design Body (Enclosure System)
15	A.17	Integrate Launch and Storage Systems	2	Xin	Xin: Integrate Launch and Storage Systems
16	A.18	Integrate Launch, Storage, and Containment Systems	5	Xin	Xin: Integrate Launch, Storage, and Containment Systems
17	TP.07	Maximum launch velocity	5	Frederic	Frederic: Maximum launch velocity
18	TP.08	Maximum mass launch capability	5	Frederic	Frederic: Maximum mass launch capability
19	A.15	Build Body (Enclosure System)	5	Frederic	Frederic: Build Body (Enclosure System)
20	A.16	Test Body (Enclosure System)	5	Frederic	Frederic: Test Body (Enclosure System)
21	A.19	Integrate Working System	5	Frederic	Frederic: Integrate Working System
22	TP.02	Test Procedure: Child fails to arm toy	5	Frederic	Frederic: Test Procedure: Child fails to arm toy.
23	TP.03	Test Procedure: Child releases armed toy near face of self or another child	5	Frederic	Frederic: Test Procedure: Child releases armed toy near face of self or another child
24	TP.04	Test Procedure: Child aims projectile at eyes of self or of another child	5	Frederic	Frederic: Test Procedure: Child aims projectile at eyes of self or of another child
25	TP.01	Test Procedure: Child plays with toy	5	Frederic	Frederic: Test Procedure: Child plays with toy
26	TP.05	Test Procedure: Child uses pet rodent as projectile	5	Frederic	Frederic: Test Procedure: Child uses pet rodent as projectile
27	TP.06	Test Procedure: Child drops or throws the toy	5	Frederic	Frederic: Test Procedure: The child drops or throws the toy.
28	TP.09	Storage volume	1	Marie	Marie: Storage volume
29	TP.10	Mean cycles to failure	5	Frederic	Frederic: Mean cycles to failure
30	TP.11	Breaking force of critical components	2	Xin	Xin: Breaking force of critical components
31	TP.12	Repair part cost	1	Marie	Marie: Repair part cost
32	TP.13	Tensile strength of home-applied adhesive bond	2	Xin	Xin: Tensile strength of home-applied adhesive bond
33	TP.14	Time for disassembly and reassembly	3	Amar	Amar: Time for disassembly and reassembly
34	TP.15	Material cost	1	Marie	Marie: Material cost
35	TP.16	Time to manufacture	3	Amar	Amar: Time to manufacture

Figure 8.6   Assigning tasks to teams.

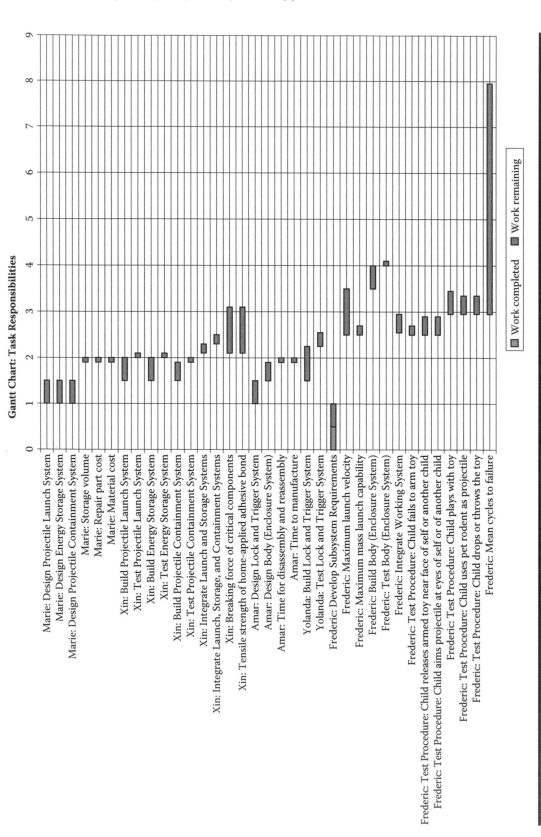

Figure 8.7 Responsibility Gantt chart with zero delays.

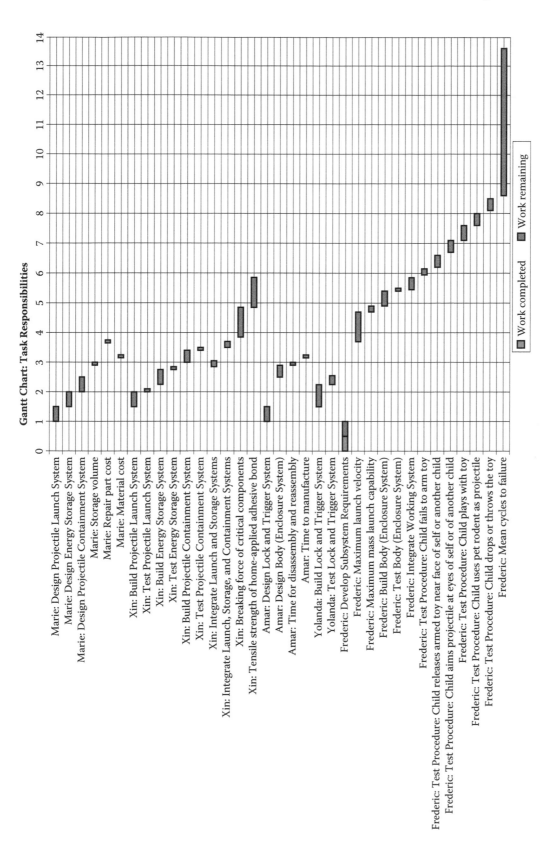

Gantt Chart: Task Responsibilities

Marie: Design Projectile Launch System
Marie: Design Energy Storage System
Marie: Design Projectile Containment System
Marie: Storage volume
Marie: Repair part cost
Marie: Material cost
Xin: Build Projectile Launch System
Xin: Test Projectile Launch System
Xin: Build Energy Storage System
Xin: Test Energy Storage System
Xin: Build Projectile Containment System
Xin: Test Projectile Containment System
Xin: Integrate Launch and Storage Systems
Xin: Breaking force of critical components
Xin: Tensile strength of home-applied adhesive bond
Amar: Design Lock and Trigger System
Amar: Design Body (Enclosure System)
Amar: Time for disassembly and reassembly
Amar: Time to manufacture
Yolanda: Build Lock and Trigger System
Yolanda: Test Lock and Trigger System
Frederic: Develop Subsystem Requirements
Frederic: Maximum launch velocity
Frederic: Maximum mass launch capability
Frederic: Build Body (Enclosure System)
Frederic: Test Body (Enclosure System)
Frederic: Integrate Working System
Frederic: Test Procedure: Child fails to arm toy
Frederic: Test Procedure: Child releases armed toy near face of self or another child
Frederic: Test Procedure: Child aims projectile at eyes of self or of another child
Frederic: Test Procedure: Child plays with toy
Frederic: Test Procedure: Child uses pet rodent as projectile
Frederic: Test Procedure: Child drops or throws the toy
Frederic: Mean cycles to failure

■ Work completed   ■ Work remaining

Figure 8.8   Responsibility Gantt chart with staggered start times.

Table 8.12 Schedule with Staggered Start Times

Task No.	Task Name	Delay Start (Qtr. Hours)	Start	Work Completed	Work Remaining	Finish
A.01	Develop subsystem requirements		0	0.5	0.5	1
A.02	Design projectile launch system	0	1	0	0.5	1.5
A.03	Build projectile launch system	0	1.5	0	0.5	2
A.04	Test projectile launch system	0	2	0	0.1	2.1
A.05	Design energy storage system	2	1.5	0	0.5	2
A.06	Build energy storage system	1	2.25	0	0.5	2.75
A.07	Test energy storage system	0	2.75	0	0.1	2.85
A.08	Design projectile containment system	4	2	0	0.5	2.5
A.09	Build projectile containment system	2	3	0	0.4	3.4
A.10	Test projectile containment system	0	3.4	0	0.1	3.5
A.11	Design lock and trigger system	0	1	0	0.5	1.5
A.12	Build lock and trigger system	0	1.5	0	0.75	2.25
A.13	Test lock and trigger system	0	2.25	0	0.3	2.55
A.14	Design body (enclosure system)	0	2.5	0	0.4	2.9
A.17	Integrate launch and storage systems	0	2.85	0	0.2	3.05
A.18	Integrate launch, storage and containment systems	0	3.5	0	0.2	3.7
TP.07	Maximum launch velocity	0	3.7	0	1	4.7
TP.08	Maximum mass launch capability	4	4.7	0	0.2	4.9
A.15	Build body (enclosure system)	0	4.9	0	0.5	5.4

Table 8.12 Schedule with Staggered Start Times (Continued)

Task No.	Task Name	Delay Start (Qtr. Hours)	Start	Work Completed	Work Remaining	Finish
A.16	Test body (enclosure system)	0	5.4	0	0.1	5.5
A.19	Integrate working system	7	5.45	0	0.4	5.85
TP.02	Test procedure: child fails to arm toy	9	5.95	0	0.2	6.15
TP.03	test procedure: child releases armed toy near face of self or of another child	10	6.2	0	0.4	6.6
TP.04	Test procedure: child aims projectile at eyes of self or of another child	12	6.7	0	0.4	7.1
TP.01	Test procedure: child plays with toy	0	7.1	0	0.5	7.6
TP.05	Test procedure: uses pet rodent as projectile	7	7.6	0	0.4	8
TP.06	Test procedure: child drops or throws the toy	9	8.1	0	0.4	8.5
TP.09	Storage volume	0	2.9	0	0.1	3
TP.10	Mean cycles to failure	11	8.6	0	5	13.6
TP.11	Breaking force of critical components	7	3.85	0	1	4.85
TP.12	Repair part cost	3	3.65	0	0.1	3.75
TP.13	Tensile strength of home-applied adhesive bond	11	4.85	0	1	5.85
TP.14	Time for disassembly and reassembly	0	2.9	0	0.1	3
TP.15	Material cost	1	3.15	0	0.1	3.25
TP.16	Time to manufacture	1	3.15	0	0.1	3.25

Observe, for example, that Frederic has nothing to do after developing the subsystem requirements until the catapult is available for the maximum launch velocity test. If Frederic is idle during this time, then we have wasted our resources. Adding more people to a project can shorten the time, but it may not be cost effective. These are trade-offs that the design project manager must face.

We have described a process for scheduling a design project with realistic consideration of precedence and resource constraints, tracking the progress of the project, and exploiting the concept of working in parallel to complete the project in less time. Specialized software packages such as MS Project can assist in this process, but we have demonstrated all the ideas using spreadsheet software.

Conduct Management Reviews

A *management review* is a technique for keeping a design project on schedule and within budget. It considers the technical aspects of a design only if these aspects affect schedule or budget. A management review typically consists of team leaders reporting progress to a project manager. To make these meetings efficient, we suggest that each team leader come to the meeting with a single piece of paper that reports his or her team's progress and status. Organize the report into four sections:

1. *Achievement.* Summarize in bullet form the major accomplishments of the team since the last management review.
2. *Schedule.* List the major activities that are planned to be accomplished before the next management review.
3. *Metrics.* Provide objective measures of activity, summarizing progress across all previous management reviews. For example, in a software project, compare the number of lines of code that have been written against a budgeted number.
4. *Issues.* List any issues that the team is wrestling with that are causing or may cause the schedule to slip or cost to exceed budget.

Table 8.13 is a management review report for the toy catapult project supposedly created by Marie, the leader of the design team. It has four quadrants, as suggested previously, and focuses on information relevant to project management.

It reports on two issues that surfaced during a customer design review. Based on Figures 6.5–6.8, one customer observed that it might be difficult for a preschooler child to reach into the enclosure in order to load the catapult. Another thought that the graphic outline of the pitcher was designed to appeal to an older child.

Observe that a collection of reports in this format allows the project manager to track the progress of each of the teams without getting inundated with technical details. The focus of attention for the meeting will naturally be on the issues that threaten schedule or budget.

When a design project runs into trouble with either schedule or cost, management action is required. Three basic courses of action are available to the project manager:

1. *Cancel the project.* This is extreme, but if the project goals cannot be met, failing to stop the project may simply result in a greater waste of resources for the same unsuccessful outcome.

Table 8.13 Single-Page, Four-Quadrant Management Review Report for Toy Catapult Project

Management Review Meeting, December 2008 *Marie, Catapult Design Leader*	
Accomplishments since last review	**Schedule**

• Completed three-dimensional sketch of product concept
  • Developed preliminary bill of materials
  • Developed preliminary reliability estimate
• Conducted design review with customer
  • Discovered dimension issue
  • Discovered appearance issue
	• (Marie) Examine access dimensions for commercial toys
	• (Marie) Make pitcher profile more cartoon-like
	  • (Marie) Add eyes and facial expression
	• (Amar) Redesign latch, latch support, and trigger arm for increased reliability
	• (Marie) Redesign enclosure for reduced cost

Metrics	**Issues**

• 23 Requirements identified
• 16 Tests specified
• 0 Requirements verified
• 0 Tests completed
	• Dimension issue from customer feedback
	  • Distance between covers limits access by preschoolers (requires skill to load)
	  • How much space between covers is required?
	  • To resolve issue, propose examination of commercial toys with similar access requirements
	• Customer feedback: Toy likely not appealing to preschoolers
	  • Propose softening profile of pitcher to make it more cartoon-like
	  • Add eyes and facial expression

2. *Allocate more manpower or resources.* Assigning more people or more highly skilled people to a task can often get it accomplished sooner. This option includes outsourcing activities that your design team is unable to complete on its own. It would also include such things as purchasing faster test equipment.

3. *Modify the goals or requirements.* If the technical challenges prove to be too great, you may need to negotiate with the owner or customer to reduce or delay either functionality or performance. This is clearly what was required in the case of the *Vasa* warship (Chapter 1). Someone needed to say "no" to the customer. The earlier the trade-offs between capability and budget can be discovered, the better. In fact, you should plan to set *tripwires*—that is, intermediate deadlines or budget expenditures—that will set off project management alarms if you cross them too late or too soon, respectively.

Summary

In this chapter we have again considered the role of the design project manager, and focused on the additional managerial responsibilities of planning and managing project execution.

We walked through the process of creating a feasible design project schedule. This included establishing the scope of the project and listing the tasks necessary to complete the project. It also included estimating task durations and tracking actual progress against those estimates. We introduced the concept of working in parallel to speed up project completion. Before considering working in parallel, it is necessary to map the input and output relationships between the tasks and to summarize these into task precedence relationships. If the tasks cannot be ordered into a sequence without precedence cycles, then the tasks must be redesigned.

Checking for cycles can be accomplished with a software macro that attempts to rearrange the precedence matrix into upper-triangular form. An earliest start schedule can be computed using formulas that can be implemented in a spreadsheet and displayed as a Gantt chart. An earliest start schedule is an extreme form of working in parallel. It considers only precedence relationships, and it ignores resource constraints. We demonstrated resource constraints by assigning each task to a particular team and requiring that delays be introduced so that no team was required to work on two tasks at the same time.

Overlaps in team schedules could be seen more easily in a Gantt chart that ordered the tasks by responsibility. The resulting scheduling problem was reduced to a computer game. Professional project scheduling software includes optimization software that searches for schedules that optimize such games.

A project schedule should be a coherent, reasonable plan for completing the tasks of a design project. Staying on schedule and within budget is best accomplished through frequent management reviews. We suggested that each team leader come to a management review with a one-page four-quadrant report listing accomplishments, schedule, metrics, and issues to keep the project manager abreast of the team's progress. The project manager has three courses of action in dealing with issues: project cancellation, resource reallocation, and requirements negotiation.

Discussions

1. Debate the wisdom of Boeing's strategy to contract with vendors to build major portions of the 787 aircraft. This practice is now commonplace in the automobile industry.
2. Relate the story of project management scope to the stories of the *Vasa* in Chapter 1 and of the Sky lawsuit in Chapter 3.

Exercises

1. Download and modify the spreadsheet "Catapult Schedule.xls" as follows. Make changes only to cells that are colored light yellow.
   a. On the sheet labeled "Teams," invent names and role descriptions for three teams. Do not use more than three teams.
   b. On the sheet labeled "Responsibility," assign each task to one of these three teams. We recommend that all tasks that involve the prototype catapult or a partial prototype be

assigned to a single team. The rule is that *two teams cannot use the same prototype or partial prototype at the same time.*

  c. On the sheet labeled "Staggered Schedule," adjust the delays to start times so that *no team is performing more than one task at a time.* Use the spinner buttons and the Gantt chart "Task Responsibilities." Use whole numbers only for the delays. You may want to drag the Gantt chart to be closer to the spinner buttons and adjust the zoom.

  d. On the sheet labeled "Duration," change the "percent complete" numbers to what you would expect them to be *4 hours* after the start of the project.

  e. (Bonus) This is a competition with other students in the class. Points will be awarded to the student who finds the shortest total time required to complete the project. Fewer points will be awarded to runner-ups. This is an exercise in finding the best way for teams to work in parallel, where "best" is measured in terms of time to project completion.

References

Boeing. The 787 Dreamliner. The Boeing Company. http://www.Boeing.com/commercial/787family/background.html (accessed May 20, 2009).

Gates, D. 2008. Chief of 787 Dreamliner's supply chain takes stress in stride. *The Seattle Times,* June 14.

Jackson, Joanna. 2009. Private communication.

Kaste, M. 2007. New cargo plane symbolizes Boeing outsourcing. *National Public Radio,* April 10.

Lewis, J. P. 2005. *Project planning, scheduling, and control: A hands-on guide to bringing projects in on time and on budget.* New York: McGraw–Hill Professional.

McInnes, I. 2008. A 787 supply chain nightmare. Aerospace-Technology.com, March 25. http://www.aerospace-technology.com/features/feature1690/ (accessed August 7, 2008).

Reuters. 2008. Boeing 787 Dreamliner may face further delays. *USA Today,* January 10.

Wayne, L., and M. Maynard. 2008. New Boeing 787 jetliner faces another delay. *The New York Times,* December 5.

Chapter 9

Iterate the Design Process

Learning is not compulsory...neither is survival.

W. Edwards Deming, 1993

Hierarchy... is one of the central structural schemes that the architect of complexity uses.

Herbert Simon, 1981

Introduction

Iterate means "to repeat." When iterated, a series of activities becomes a cycle of activities. There are four senses in which we can use the concept of cycles and *iteration* in extending the process of Getting Design Right:

1. Iterate until feasible (backtrack).
2. Iterate with improvement.
3. Iterate by level.
4. Dive and surface.

Iterate until Feasible

Up until this point, we have presented the Getting Design Right methodology primarily as a linear process ("a place to start and steps to follow") with definable results. What happens when things go wrong? What happens when you discover that the design concept cannot possibly satisfy all of the system requirements? What happens if it is not possible to complete the project in the time allowed with the current budget? What happens when a module or a prototype fails a test? What happens when different design teams have been assigned tasks that cannot be performed independently of each other?

In these cases, we have hinted that some form of *backtracking* or iteration is required. In fact, backtracking is an essential part of any problem-solving methodology. It is vain to hope that you can conduct any design effort without backtracking. On the other hand, backtracking usually means you have wasted time and money on a failed concept, so the Getting Design Right process has emphasized steps you can take to reduce the chance of backtracking. Of these, collecting a clear set of initial requirements and exploring the design space are the most important.

Hitting a Dead End: Propulsion Module for the International Space Station

The National Aeronautics and Space Administration (NASA) faced many challenges in developing and supporting the International Space Station [Figure 9.1]. These challenges, such as Russian difficulty in completing its components on schedule, led NASA to pursue development of a U.S. propulsion capability for the space station to serve as an alternative to the planned Russian capability. In 1998, NASA accepted a proposal from Boeing Reusable Space Systems for a U.S. propulsion module. NASA's initial effort to develop this module was not successful in meeting the program's performance, cost, and schedule goals. The effort failed to produce a design that met mission requirements, increased its estimated cost by $265 million (from $479 to $744 million), and slipped its schedule by about 2 years. NASA eventually canceled the program and initiated a follow-on effort...

The initial propulsion module project did not meet performance, cost, and schedule goals largely because NASA proceeded with Boeing's proposal without following fundamental processes involving project planning and execution. NASA officials stated that, had these processes been followed, they would have determined earlier in the program that the Boeing proposal would not meet project goals. For example, NASA did not complete a project plan or develop sufficient information in areas such as systems analysis and risk management to guide the program. Having such basic information is fundamental to sound project management. In addition, Boeing's design was accepted and implemented before the propulsion module's detailed technical requirements were fully established. NASA later found that the design was not as mature as anticipated and that it required substantial changes. This led to significant delays, cost increases, and, ultimately, project cancellation.

In May 2000, NASA began to assess alternatives to the Boeing-proposed propulsion module. The assessment team defined mission success criteria, identified key design assumptions, and performed comparative analyses on competing designs. Based on its analyses, the team recommended a follow-on design. According to NASA officials, this effort brought early analytical rigor to requirements definition, which NASA had failed to do in the initial project.

NASA acknowledged that its initial approach to developing a propulsion module was inadequate and contributed to the project's unsuccessful conclusion. NASA officials performed lessons-learned efforts on the project in general and on one specific component—on-orbit fuel transfer—hoping to avoid similar problems in managing future programs. In all cases, NASA concluded that the lack of an early systems analysis contributed to project failure. Regarding the failed attempt to design an on-orbit fuel transfer component into the propulsion module, NASA cited difficulty in establishing requirements, estimating cost and schedule, and providing adequate resources.

U.S. General Accounting Office Report (2001)

Figure 9.1 International Space Station. (Source: NASA.)

Backtracking Strategies

If a problem does arise, consider all of the following backtracking strategies:

1. Cancel the project or change the mission.
2. Revise the requirement, reduce the scope, or increase the budget. Negotiating the scope of the design project is a legitimate activity.
3. Relax the exit criteria. If the module or prototype cannot pass the test, then perhaps the exit criteria are too stringent. Do exit criteria exist that the module or prototype could pass? Are these criteria acceptable to the user, the customer, or the next higher design organization? This approach has the stigma of being "the lazy man's solution" to passing a test. Sometimes, however, exit criteria can be somewhat arbitrary. It is worth discovering the true requirement.
4. Improve the components or subsystems. The problem may not lie with the design: The problem may be due to failures in components or lower level subsystems.
5. Adjust the design parameters. Perhaps the concept is good but the parameters of the design are not optimized. Systematically explore the parameter space to find the best combination of parameters.
6. Choose another design concept. Return to the concept selection matrix and pick the next best alternative design concept. Add a row to the matrix indicating the likelihood that the integrated concept will pass the test failed by the current concept. Evaluate each integrated design concept against this additional attribute.

7. Backtrack through the concept classification tree. Is there another way to perform this function? Revisit the concept classification tree. You may have to back up several levels and try a different branch.

8. Combine functions. Sometimes a single part can replace a collection of others by performing more than one function.

9. State the function or the problem more abstractly. This might be equivalent to introducing a new branch point on the concept classification tree. On the other hand, perhaps you have been too restrictive in how you have named the problem.

10. Introduce a new function or task or a new parameter. Sometimes there are not enough design degrees of freedom to satisfy all of the requirements. Dividing up the function or the task into finer detail may reveal new degrees of freedom.

Tables 9.1 and 9.2 illustrate these concepts using a variety of problems that might arise in the toy catapult project. For each problem, we think of solutions (not all of them serious) that the different strategies might suggest. Some solutions are obviously better than others. For example, increasing the launch velocity to pass TP.07 should be a simple matter of adjusting the design parameters. Finding an example of combining functions (function sharing) was not difficult. As shown in Figure 9.2, a removable pin can serve both to secure the catapult into position and to detect the command to release. That is, it may be possible to design a pin with a notch in it that would hold the catapult arm in position but, at the same time, be easy to knock out from any direction. This is only a concept fragment. Satisfying the function "detect receptacle in position" may require the inclusion of a spring to return it to the arming position.

The different backtracking strategies are used to suggest possible solutions to the problem. Use Pugh analysis (Chapter 5) to select the solution concept that best solves the problem. Any change to the design may ripple through and require changes in other subsystems.

Iterate with Improvement

For many years, the purpose of inspection in manufacturing was to detect and remove parts that did not meet specifications. The philosophy of modern quality control is quite different. Under the philosophy of *continuous improvement,* every production cycle is an opportunity to learn and improve. The inspection part of the production cycle provides the data used to measure improvement. Table 9.3 describes the *plan–do–check–act* (PDCA) cycle that expresses the essence of the continuous improvement philosophy. It is also known as the PDSA cycle (variously interpreted as "plan–do–study–act" and "plan–do–see–approve").

The PDSA cycle was popularized by W. Edwards Deming in the 1950s. Deming called it "the Shewhart cycle" in honor of his colleague and collaborator, Walter A. Shewhart. This philosophy was adopted and used aggressively in the Japanese automobile industry, most particularly by Toyota. From there, it has found its way into every manufacturing industry worldwide, and its impact is felt even in nonmanufacturing settings such as health care, information systems, business processes, and educational services.

The Eight Steps to Getting Design Right process is a derivative of the PDCA cycle. Figure 9.3 displays the PDCA cycle together with the Getting Design Right cycle. We developed the Eight Steps to Getting Design Right nomenclature to adapt the PDCA cycle to problem solving in a design context. Similar cycles can be found in the "Design for Six Sigma" literature and in

Table 9.1 Possible Backtracking Solutions to Toy Catapult Problems

Strategy no.	Strategy	Problem				
		Prototype Fails TP.07 (Maximum Launch Velocity)	Prototype Fails TP.05 (No Harm to Pet Rodent)	Design Fails TP.15 (Material Cost)	Energy Storage System Cannot Be Designed Independently of Projectile Launch and Projectile Containment Systems	Crossbeam Fails TP.11, Breaking Force
1	Cancel the project or change the mission.		Abandon toy catapult concept: Safety concerns outweigh fun; make a wooden puzzle instead.			
2	Revise the requirement, reduce the scope, or increase the budget.	Reduce maximum velocity requirement; 4 mps may be fast enough for fun factor.	Reduce maximum velocity requirement and maximum mass launch capability.	Increase the material cost budget.		
3	Relax the exit criteria.	It is sufficient if in 1 trial out of 50, launch velocity reaches 6 mps.	Banana used for test does not have to be ripe.			
4	Improve components or subsystems.					Choose different material type or vendor.
5	Adjust the design parameters.	Increase length of catapult arm; increase mass of pendulum weight.		Use thinner enclosure material.		Increase dimensions of crossbeam.

Table 9.2 Possible Backtracking Solutions to Toy Catapult Problems

No.	Strategy	Problem			
		Prototype Fails TP.07 (Maximum Launch Velocity)	Design Fails TP.15 (Material Cost)	Energy Storage System Cannot Be Designed Independently of Projectile Launch and Projectile Containment Systems	Crossbeam Fails TP.11, Breaking Force
6	Choose another design concept.	Choose spring and catapult concept.			
7	Backtrack through the concept classification tree.	Reconsider chemical concept fragment.			
8	Combine functions.		Eliminate latch mechanism from lock and trigger system: Use removable pin both to catch catapult arm and to act as trigger.	Assign all design tasks for these three subsystems to a single design team.	
9	State the function more abstractly.				Projectile containment does not need to be a rigid receptacle; hard stop is not needed. Eliminate vibrations by using a sling to hold and release the projectile.
10	Introduce a new function or task or a new parameter.	Add spring to receptacle attachment. Hard stop causes spring to transmit additional energy to projectile.		Include a new task to set major design parameters (pendulum mass, catapult arm length, receptacle mass, pivot hole size, attachment dimensions) for all three subsystems and then leave residual design decisions (pendulum weight dimensions, material density, catapult arm material selection, receptacle design, etc.) in original subsystem design tasks.	

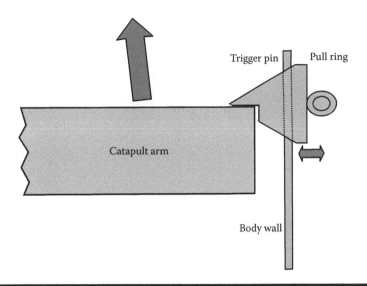

Figure 9.2 Trigger pin concept eliminates latch mechanism.

the systems engineering literature. The Eight Steps to Getting Design Right process is an integration of these two complementary schools of thought (Design for Six Sigma and systems engineering).

The basic cycle of Getting Design Right is define–measure–explore–optimize–design–validate. This corresponds roughly to the PDC (plan–do–check) portion of the PDCA cycle. Note that we have broken out the measurement step as an explicit part of the up-front process. Identifying the metrics by which we will judge success (i.e., the "measures of effectiveness") is an important part of the design cycle, and it needs to be done early. Benchmarking the competition was included in this step. As we described it, the focus of the validation step was internal to the design cycle, but it could include ongoing capture of the measures of effectiveness to determine the success of the design. For example, we could include a study of how our grandchildren actually play with the toy catapult, as described in Chapter 3 "Measure the Need and Set Targets."

In general, the data collected during the validate step of one design cycle can be used in the measure step of the next design cycle. Seen in this light, the execute and iterate steps of the Eight

Table 9.3 The Plan–Do–Check–Act (PDCA) Cycle of Continuous Improvement

Plan	Identify an opportunity for improvement; establish a baseline of current performance and set a measurable target for improvement.
Do	Implement the change, preferably on a small scale, and measure the change in performance.
Check	Analyze the change, measuring its improvement against the baseline and the target, and then publish the results.
Act	Institutionalize the change, if successful, modifying the process to include the change permanently.

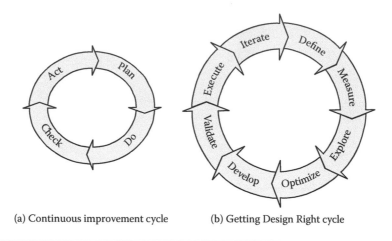

(a) Continuous improvement cycle   (b) Getting Design Right cycle

Figure 9.3   Continuous improvement and Getting Design Right cycles.

Steps to Getting Design Right match the "act" step of the PDCA cycle. Each time that we engage in design is an opportunity to learn and improve, provided we take the time to measure success and to relate that success, or lack of it, back to the specific processes we used for design.

The improvement activity is very much in your hands as you engage in design. In the introduction, we pointed out that you will be involved in many design cycles over the course of your life and career. Not all of them will lead to successes, but every one of them is an opportunity to learn and improve. We have suggested, along the way, steps you can take to develop your skill and experience. To recap, these steps include the following:

- *Tailor the process.* This means, first of all, to recognize that you are entering a design cycle. If you tackle a project without recognizing its design nature, you are likely to skip all of the up-front steps and end by implementing an inferior solution. Second, it means to be deliberative about how you will conduct the activity: what steps you will follow and to what level of detail you will go.
- *Establish measures of effectiveness and benchmark the competition.* These are ways to provide a quantitative basis for judging your success.
- *Practice writing and testing requirements.* Discovering the difficulty of testing requirements is the best way to improve your skill at writing them. Writing sharper, more focused requirements is the best way to achieve improvements in design. For example, a requirement such as "the software user interface shall be user friendly" is not testable and is unlikely to result in an effective design. Detailed requirements, such as "the user shall be able to save the current workspace using a single mouse-click," are testable, and they are more likely to lead to effective designs.
- *Track your performance.* How long did the different activities in your work breakdown structure take? Knowing the time you spent on certain activities in the past will allow you to predict their durations better in the future.
- *Reflect and improve.* Take the time with your project team to debrief the results of the design cycle. What went wrong? What went right? Why? On what did you spend too much time? Where in the process did you not spend enough time? What could be improved on

the next design cycle? As a project manager, you should factor that debriefing time into the work breakdown schedule. It should be a budgeted activity.

Writing this book has been an exercise in design. We seek your feedback on how well we have achieved the objectives identified in the introduction. All feedback is welcome. An adage in design goes, "Even the phrase 'I hate it!' is direction." Having direction means that improvement is possible.

Iterate by Level

The iterate step in the Eight Steps to Getting Design Right process can also be interpreted in the sense of "iterate by level." Rather than backtracking within a design cycle or from one product design cycle to the next, this is a nested form of iteration. The design cycle is interrupted while a more detailed design cycle takes over within it, and then it continues once the nested cycle has finished.

Level-by-Level Decomposition: The Vee Diagram

A purposeful system can be understood in terms of its context, its behavior, and its subsystems. Complex systems are typically designed in layers, and each layer is viewed as a system. Hence, each layer has a context, behavior, and subsystems. Each subsystem can also be viewed as a system with context, behavior, and subsystems, called sub-subsystems. The process of understanding a complex system continues in this top-down fashion until you encounter a layer or subsystems that can be designed and built using known components and techniques. This top-down process for complex systems design is typically called the *systems engineering process.*

The power of the Getting Design Right process comes from its applicability to any level of design activity. The systems engineering process can be viewed as simply the application of the Eight Steps to Getting Design Right process to every layer of a complex system and to every subsystem of every layer.

The "Vee diagram" (Fossberg and Mooz 1992) is a classic expression of the level-by-level decomposition of a complex system and the process of design, build, and test to develop and deploy the system. Figure 9.4 is a variation of the Vee diagram emphasizing our cyclic view of the process. The left-hand side of the V is the process of developing requirements, identifying subsystems, detailing the functions of these subsystems, allocating performance requirements, and specifying interfaces. This process is repeated for each subsystem, each sub-subsystem, and so on, until a point is reached where detailed specifications can be written.

The design–build portion at the base of the V is where traditional engineering takes over and engineering domain experts develop particular solutions to meet the specifications. The right-hand side of the V describes the long process of validating the design. Design validation begins with tests to verify that components satisfy the requirements set for them. As tests of components and modules are completed, the system is gradually assembled and integrated. Tests are designed at each level of integration to verify the requirements set for that level. Ultimately, user and market tests validate whether the system as designed and built fulfills the design mission.

Figure 9.4 is an idealized picture of the design process. It does not capture the backtracking that takes place in complex systems design and does not capture the parallel nature of many of these activities. For example, software engineers can be developing building blocks of code without knowing exactly how these building blocks will be integrated into the final product.

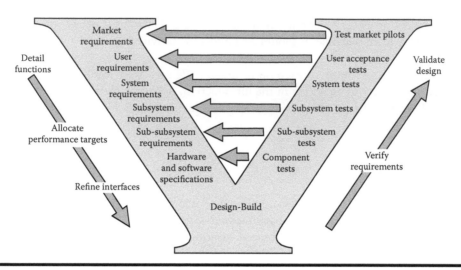

Figure 9.4 The "Vee" diagram.

In other words, a layer does not need to be specified completely before design, build, and test work can begin on lower layers. Even in these situations, however, it should be possible to trace low-level requirements to high-level system purposes. We illustrated this notion of levels with the toy catapult, identifying its various subsystems and planning the verification of its design requirements in a staged manner (component test, partial prototype tests, and finished prototype tests).

Three features of the systems engineering process are worth emphasizing:

- detailing functions into behaviors;
- allocating nonbehavioral requirements; and
- maintaining hierarchies and traceabilities.

Detail Functions into Behaviors

In Chapter 6, "Develop the Architecture," we showed how a use case behavior could be described in terms of different subsystems performing functions according to a thread of behavior, with interfaces transferring messages, material, or energy from one subsystem to another to execute the behavior. The operational description template was a unifying approach to describing behavior, subsystem functional responsibilities, and interface requirements. In a complex system, subsystems are abstract collections of more fundamental objects. The subsystem function is really achieved by these more fundamental objects working together. What is a function at the level of the subsystem is really a behavior within the subsystem of the sub-subsystems working together.

The systems engineering process, therefore, requires that each function be described by a behavior of functions allocated to a lower level of subsystems (with interfaces, states, triggers, and events) and the process repeated for each lower level function until a level of subsystem is reached that can be designed, built, and tested by an engineering organization. This process is often called "functional decomposition," but this phrase misses the crucial concept of behavior. We prefer the phrase "detail functions into behaviors."

Allocate Nonbehavioral Requirements: Linked Houses of Quality

Just as behavioral requirements are detailed into functions and allocated to subsystems (becoming behaviors detailed and allocated to subsystems, sub-subsystems, and so on), so must nonbehavioral requirements be allocated and detailed to lower levels. Some engineering characteristics, such as mass, exist at every level for every physical system and component. Consequently, a high-level requirement such as "the mass of the vehicle shall not exceed 2,000 kg" must be allocated across subsystems at each level. We illustrated the process of requirement allocation in Chapter 6 for cost and reliability measures.

For other nonbehavioral requirements, the performance measure at one level of design is really the result of interacting characteristics among its subsystems. We illustrated this concept in Chapter 5, "Optimize the Design," with a linked house of quality (Figure 5.3) that related the system-level engineering characteristics of launch mass and velocity with subsystem-level engineering characteristics such as pendulum counterweight mass, length of arm, and location of pivot. Some form of optimization or trade-off analysis is required to set target requirements for the lower level engineering characteristics. This led us to the formulation of optimization problems for continuous design parameters.

A conceptual framework for this allocation process is the concept of *linked houses of quality*. Figure 9.5 illustrates the concept by showing a family of three houses of quality. The first house, in the upper left, describes how customer attributes ("customer requirements") are considered when setting targets for various engineering characteristics of the system ("system characteristics"). The

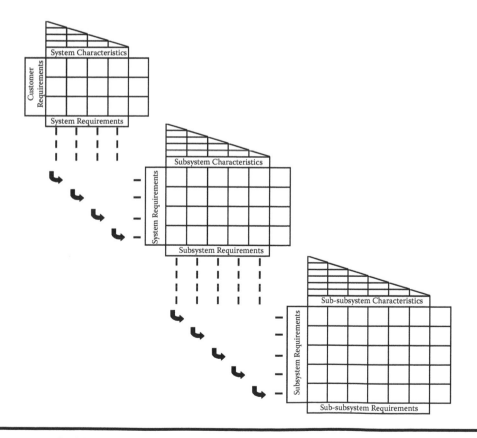

Figure 9.5   Linked houses of quality.

back porch of the house is used for benchmarking and the roof (displayed here as a shed roof, a lower triangular matrix) captures the engineering relationships among the system characteristics. The basement is where targets for the system characteristics are recorded. At this point, these targets have become system requirements.

The next house represents the subsystems. Now, however, the rows of the main floor of the house correspond to the system requirements that emerged from the first house of quality. The arrows in the figure suggest the transposition of columns from one house into rows of the next house. This second house describes how targets are set for characteristics of the subsystems. In this case, however, there is likely to be considerable engineering knowledge. The main room of the house captures engineering relationships (preferably in the form of mathematical models) between the system characteristics (rows) and the subsystem characteristics (columns). Recall the "imputed importance" row of the house of quality. That row would conceptually become the "relative importance" column of the front porch. The roof of this house captures engineering relationships and constraints that exist among the subsystem characteristics. The back porch can continue to represent the area for competitive benchmarking, and the basement represents the setting of subsystem targets (now called "subsystem requirements") to achieve all of the system requirements.

The third house of quality, in the lower right, repeats the concept for the next lower level. The increase in size of each successive house suggests the greater level of detail that must occur as we move to lower levels. We have shown only three houses, but, conceptually, there is a house for every level, for every subsystem, and for every process (including manufacturing, installation, operation, maintenance, and disposal). It is a grand concept. The goal in many companies is to develop analytical computer models that capture the engineering relationships in all of these potential houses of quality. The integration of all these models would then allow the company to have a view of design that is completely specified by design parameters, a so-called *fully parameterized design*.

Maintain Hierarchies and Traceabilities

For the toy catapult project, we have been able to use simple spreadsheets to describe the system and all of the relationships. For a complex system, databases are required to keep track of all the relationships more efficiently. The following is a partial listing of the data found in systems engineering databases:

- *requirements traceability* (how requirements at one level are traceable to requirements at the next higher level);
- bill of materials (subsystems, sub-subsystems, modules, components, etc.);
- behaviors to functions, functions into behaviors;
- functions assigned to objects (subsystems, sub-subsystems, modules, components, etc.);
- physical links between subsystems;
- data and data flow;
- test plan (by level);
- work breakdown structure (by level); and
- project schedule (by level).

Many of these are *hierarchical relationships*. A set of relationships is hierarchical if it can be drawn as a tree in which the trunk divides into branches and the branches into smaller branches and so on until the branches become twigs and the twigs bear leaves. Each object can be associated with a unique joining point, or *parent*. For example, it should be possible to trace every component through the bill of materials to find the module to which it belongs, the sub-subsystem to which it belongs, and so on up the levels). Similarly, low-level functions should be traceable upward to high-level functions. The work breakdown structure is hierarchical. However, because of the complexity implicit in the linked houses of quality concept, it may not be possible to draw the requirements relationships as a tree; that is, requirements relationships may not be hierarchical. Nevertheless, it should be possible to take any low-level requirement and trace it back upward to a collection of originating requirements.

Dive and Surface: A Systems View

We complete this chapter and this book by considering one more interpretation of cycles and iteration within the Getting Design Right process. We have made frequent use of the dive-and-surface paradigm. Diving is the act of considering a problem at a detailed level. Surfacing is the act of summarizing the details. It is during the surfacing phase that we develop more of a systems view of a problem or situation. Sometimes we dive simply to make sure that our systems view is more realistic. We associate the affinity process with surfacing. The affinity process can be applied any time we have a collection of related but nonhomogeneous details.

By the end of any design cycle, you are likely faced with a whole collection of odd things you have created. After several design cycles, you will start to see similarities and affinities among these odd things. A surfacing phase in which you collect these odd things into groups will help to suggest patterns, tools, templates, and processes that would unify these odd things and make it easy to create similar things in the future. The development of this text, for example, has led to the creation of reusable spreadsheet templates for quality function deployment, systems design, and project scheduling.

It is fitting to close this book with a surfacing example that will also serve as a review of the concepts we have covered. Looking at the index at the end of this book, you can find a list of all of the major concepts covered in this book, organized alphabetically. It is a collection of related but nonhomogeneous details. Our final exercise is to apply the affinity process to this collection of concepts.

Rather than begin the affinity process cold, without any conception of the grouping columns, we look for an organizing principle. John Zachman (1987) published a seminal article on information systems design in which he recommends approaching a systems design framework with the questions "who, what, where, when, how, and why" against a series of levels or domains. Table 9.4 is a simplification and adaptation of his generic framework. We have extended the columns of his framework to include the question "how well" and we have simplified the number of levels.

The exercise we conducted was to drag items from the index of this book and drop them into one or more of the cells in Table 9.4. These were subjective judgments, and they may not agree with how you would classify the concepts. Tables 9.5–9.7 show the resulting affinity matrices.

Several observations jump out as a result of the affinity exercise. The first is that we have written very little in this book that would fall into the "where" category, even though we can think of many ways in which this category is important in design and engineering. In physical designs,

Table 9.4 Affinity Categories for Getting Design Right

	Who	What	Where	When	How	Why	How Well
Context							
System							
Designer							

Table 9.5 Affinity Matrix: System Context

		Who	What	Where	When	How	Why	How Well
System Context		client	context diagram		market growth estimate	secondary use case behavior	market requirements specification document (MRS)	market share estimate
		customer	context matrix			failure modes	value proposition	market size estimate
		competition	external entities				customer comments	benchmarks
		unintended users	system boundary				secondary use cases	risk priority number (RPN)
							emergent behavior (gridlock or infinite loop)	severity
							emergent interactions	likelihood of causal occurrence
							risk (chance of negative outcome)	

three-dimensional spatial layouts are critical for determining whether parts will fit together or whether clearance around a part is sufficient to permit its free movement. Communications networks, manufacturing supply chains, and transportation systems are examples of systems in which geographical locations and relationships are critical.

Another gap that shows up in the affinity matrix is that we have not described anything that relates to how well the design organization ("the designer") performs its mission. The "how well" column in Table 9.7 is empty. Here, again, this is not because the category is unimportant, but rather because of our focus. Design organizations use a variety of metrics to evaluate their performance, including "percentage of projects completed on time and on budget," "number of projects cancelled," "number of derivative products created," "number of patent applications," and so on.

The second step in the affinity process is to summarize the categories. We do that separately for the rows and columns of the affinity matrix. Table 9.8 is a summary of the rows of the affinity matrix. We have chosen to refer to these as "domains." There is a fairly clear distinction between the design organization and the system being designed. It is also useful to treat the system context as a separate domain.

Table 9.9 summarizes the columns of the affinity matrix. Observe that we have reordered the columns to match the emphasis they received within the Eight Steps to Getting Design Right process more closely. We considered "who," "why," and "how" early in the process (define the problem). Then we considered "how well" (measure the need). Later we considered "what" and "when." The category of "where" came more as an afterthought. We place that column last. We refer to these different columns as "views" and we further divide a view into its entities and its relationships ("from–to"). Table 9.9

Table 9.6   Affinity Matrix: System

	Who	What	Where	When	How	Why	How Well
	user	external entities		operational description template (ODT)	behavior	purposeful system	goal–question–metric (GQM) matrix
		system boundary		functional view	thread	use case	measure of effectiveness
		product family		precedence matrix	stimulus	primary capability	house of quality
		system		infinite loop	response	product objectives	engineering characteristics
		subsystem		gridlock	operational description template (ODT)	house of quality	technical performance measures
System		component		simple cycle	test thread specifications	imputed importance	system reliability
		bill of materials		event	functional requirements	originating requirements	mean time to failure (MTTF)
		operator interface specifications			function names	requirements trace matrix	failure rate
		hardware requirements specifications			functional interrelationship matrix	nonbehavioral requirements	test plan
		software requirements specifications			functional flow block diagram	target technical performance measures	exit condition
		interface requirements specifications			state change diagram		statistical acceptance test

(Continued)

Table 9.6   Affinity Matrix: System (Continued)

Who	What	Where	When	How	Why	How Well
	interface rows			state change matrix	emergent interactions	verification cross-reference matrix (VCRM)
	link			state change view	constraint	
	interface matrix			state of the system		
	design structure matrix			process plan		
	interaction matrix					

Table 9.7   Affinity Matrix: Designer

	Who	What	Where	When	How	Why	How Well
	owner	concept sketch	test facilities	thrashing	systems approach	mission statement	
	architect	concept fragment		circular discussions	Eight Steps to Getting Design Right	Pugh matrix	
	builder	concept classification tree		work breakdown structure	diving and surfacing	trade study	
	design project manager	concept combination table		task duration	tailoring the process	design risk	
Designer	design teams	morphology box		percent completed	contextual inquiry	corrective action	
	anthropologist (contextual inquiry)	integrated concept		working in parallel	affinity process	risk mitigation strategy	
	apprentice (contextual inquiry)	reference concept		task precedence matrix	Goal–question–metric (GQM) method	design risk management	
	partner (contextual inquiry)	deliverable		entry condition	Analytic hierarchy process		
		drawings		Gantt chart	Pugh analysis		
		work breakdown structure		schedule	brainstorming		

(Continued)

Table 9.7 Affinity Matrix: Designer (Continued)

Who	What	Where	When	How	Why	How well
	process plan		crashing a schedule	cluster the subsystems		
			earliest start schedule	task input–output matrix		
			resource constraints	verification and validation (V&V)		
			staggered start times	"A, I, D, T"		
				failure modes and effects analysis (FMEA)		

Table 9.8 Domains of Getting Design Right

	Domain
Context	Market opportunity, competition, the value proposition, external entities, system boundary and interfaces with other entities, stakeholders, unintended users, malicious users, threats, risks, strategy, and emergent behavior
System	User and use cases, system requirements (behavioral and nonbehavioral), physical architecture, functional architecture, control architecture, design specifications, and system validation
Designer	The design organization: its mission, roles, responsibilities, resources, learning and exploration processes, valuation and decision-making processes, verification processes, project management, and risk management

Table 9.9 Views of Getting Design Right

	Who	Why	How	How well	What	When	Where
Sample entities	Roles, perspectives, points of view	Purpose, mission, goals, objectives, values, uses, constraints, requirements, risks	Behaviors, processes, functions, states, tasks	Measures of effectiveness, business and technical performance measures, tests and benchmarks, chance and negative outcomes	Artifacts, objects, data, properties	Timing, events, tasks, durations, schedules	Facilities, geographical or spatial locations, infrastructure
Sample relationships	Organization, responsibilities, reporting relationships	Trade-offs, priorities, uses to behaviors, ends to means, conditions to actions (rules), requirements to requirements, requirements to functions, risks to strategies	Functions to objects, sequence, iteration, triggers, functional decomposition, inputs and outputs	Measures of effectiveness to engineering characteristics, tests to requirements, allocation of targets to subsystems	Physical architecture, bill of materials, interactions, interfaces, messages	State transitions, precedence, tasks to resources, conflicts, gridlock, feedback control	Networks, flow, distances and clearances

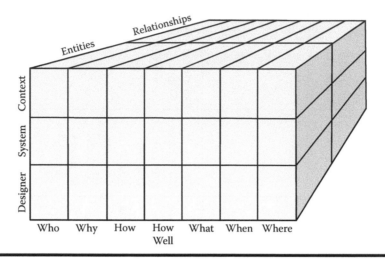

Figure 9.6   The systems cube: domains and views for Getting Design Right.

provides examples of each. You can imagine displaying each of the relationships in either a tabular (matrix) view or as some form of network or sequence graph.

Even though we spent most of the book considering a simple child's toy, we have ended with the ability to think about the design of complex systems using a robust framework. If we put Table 9.8 together with Table 9.9, we can conceive of Getting Design Right in terms of a three-dimensional matrix of concepts (Figure 9.6). We refer to this matrix as the *systems cube*. Every cell of this matrix suggests a domain, a view, and a listing of either entities or relationships. By combining the descriptions of Table 9.8 with Table 9.9, we can systematically identify the information that is relevant to that particular domain view.

Of what use is such a cube? First of all, it is a completeness check. It reveals, for example, that this book is incomplete: We do not cover the "where" view. One use of such a cube would be to classify textbooks, training seminars, and software packages according to which cells they cover. There is no one software package, for example, that claims to support completely every cell of this cube. In general, you will need to use several packages. Another use of the cube is to organize new information that you acquire. As you learn more about systems and systems design, the systems cube gives you a place to put those concepts. The Getting Design Right cycle (Figure 9.3), the "Vee" diagram (Figure 9.4), the linked houses of quality (Figure 9.5), and the systems cube (Figure 9.6) are extremely useful mental models of the design process for complex systems.

Diving and surfacing are fundamental to systems thinking. By practicing the surfacing phase both within and after every design cycle, you will grow your ability to think holistically. Without this effort, you risk repeating each cycle with memory but without understanding.

Summary

The final step of the Getting Design Right process is to "iterate the design." The concept of cycles and iterations has several interpretations, each of which is valuable.

One view of iteration is to iterate a problem-solving methodology until a feasible solution is found. This means backtracking through the process whenever requirements for a feasible solution are found to be in conflict. We suggested 10 basic strategies to use when backtracking and illustrated their use within the context of the toy catapult example.

Another view of iteration is from one design cycle to the next. The philosophy of continuous improvement looks upon each cycle as an opportunity for learning and improvement. Effective continuous improvement requires conscientious measurement and capture of information, as well as deliberate reflection and improvement. We related the Eight Steps to Getting Design Right process to the plan–do–check–act cycle of continuous improvement.

A third view of iteration is of a nested iteration of the basic design process down through the levels of a complex system design, with verification and validation coming up through the levels (the "Vee" diagram). This is referred to as the systems engineering process. This process involves, at each level, detailing functions into behaviors, allocating nonbehavioral requirements, and maintaining hierarchies and traceabilities. Database software is useful in tracking all of the relationships.

Finally, a fourth view of iteration is to complete every design cycle with a surfacing phase to improve your ability to conceptualize at higher levels of systems thinking. We illustrated the approach by applying the affinity process to the concepts of this book. This led to a generic systems cube framework for thinking about design.

As a final summary of the text as a whole, we include Figure 9.7, an annotated view of the Eight Steps of the Getting Design Right process. Each of the major concepts introduced in the

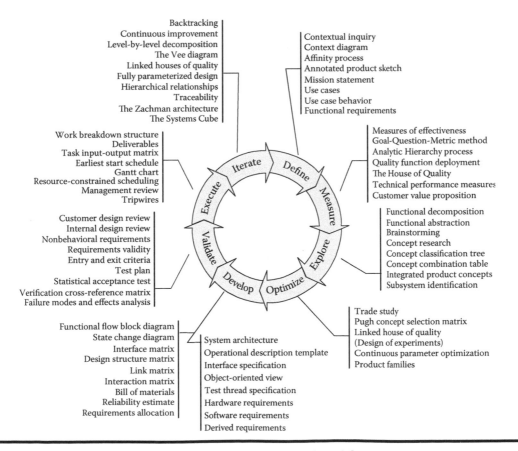

Figure 9.7 Annotated view of Eight Steps to Getting Design Right.

text is mapped to one of the eight steps. Use this figure as a handy reference to find the chapter where each of these concepts is introduced.

What Is Next?

The Getting Design Right process is an essential approach to design in the sense that we have distilled the essence of design from a number of different but related approaches to design, engineering, process improvement, product development, systems engineering, and general problem solving. It is a practical approach that is immediately applicable in your career and personal life. It remains for you to try it, experience its benefits, and improve upon it.

Of course, there is more to learn. Numerous additional tools and techniques exist. Software packages are available. Case studies of design and design failure can be found. There are also more advanced concepts, philosophical discussions, and mathematical models. Welcome to systems thinking!

Discussions

1. Is backtracking intrinsic to design? Discuss this statement in relation to the ISS propulsion module.
2. Discuss the role of interfaces in complex systems. The Internet is facilitated by a collection of communication protocols—that is, interfaces—that govern how information is transferred from one computer network to another. How much of the entire system (online commerce, social networking, music sharing, distributed computing, etc.) was needed in view when designing the interfaces?

References

Deming, W. E. 1993. *The new economics: For industry, government, education.* Cambridge, MA: MIT Cambridge Center for Advanced Engineering Study.

Fossberg, K., and H. Mooz. 1992. The relationship of systems engineering to the project cycle. *Engineering Management Journal* 4 (3): 36–43.

NASA. Photograph of International Space Station. http://www.satobs.org/image/isssts100.jpg (accessed December 22, 2008).

Simon, H. 1981. *The sciences of the artificial,* 2nd ed. Cambridge, MA: MIT Press.

U.S. GAO. 2001. Space station: Inadequate planning and design led to propulsion module project failure. United States General Accounting Office Report GAO-01-633.

Zachman, J. A. 1987. A framework for information systems architecture. *IBM Systems Journal* 26 (3): 276–292.

Appendix A: Case Studies

In this appendix we present examples of real design organizations and their activities. It should be easy to relate these activities to different steps in the Getting Design Right process. This appendix is adapted from lectures by Professor Al George, the J. F. Carr Professor of Mechanical Engineering at Cornell University. Professor George is the faculty advisor to the Cornell Formula SAE team and the founding director of the Cornell Systems Engineering Program.

Case Study: The Harley-Davidson Motor Company

What follows is a personal view of the Harley-Davidson Motor Company by Professor George. In addition to being a Harley-Davidson motorcycle owner and enthusiast, Professor George spent a sabbatical leave working at Harley-Davidson observing their practices and conducting research in engine design, among other activities.

Company Background

Harley-Davidson (with subsidiary Buell) is the only major domestic producer of motorcycles in the United States, but it has many strong overseas competitors. In spite of fierce competition, it has maintained a market share lead (49.5%) of the heavyweight motorcycle (651+cc) market. Its consolidated revenues for 2004 exceeded $5 billion (Harley-Davidson Annual Report 2004).

Traditional engineering metrics of performance, cost, and time do not explain Harley's success. In fact, in spite of a focus on these metrics for many years, Harley was near bankruptcy in the early 1980s. What turned the company around was a renewed focus on the customer and a more human-centered engineering approach. It recovered from near-bankruptcy by paying attention to quality, styling, and "customer lifestyle."

The Motorcycle Marketplace

Motorcycles are designed for a wide variety of purposes (touring, racing, and street use) and environments (road, track, and dirt). They have a wide range of appeal to people of different ages, income, interests, and lifestyles. As a result, there are numerous types of styles, performance, and comfort levels. They range in cost from several thousand to tens of thousands of dollars.

One very important phenomenon in the motorcycle marketplace is the passion of many consumers for custom-built motorcycles. Only a minority of bike owners pursues this passion, but their creations work to define the culture and influence the imagination of the larger biking communities.

Steps to Learning Customer Wishes

Harley-Davidson has a very conscious program of contextual inquiry. For them, contextual inquiry takes the following steps.

Encourage Employees to Become Active Motorcyclists

Harley-Davidson encourages its employees to use the company products and to adopt some of the various motorcycle lifestyles. Some Harley employees are every bit as big and bushy as some of the physical extremes associated with the hard-edged motorcycle world. Furthermore, employees are given significant discounts on bikes, accessories, clothing, and gifts.

Attend Rallies on Company Time

All Harley-Davidson employees, including the chief executive officer, vice presidents, engineers, and mechanics, are encouraged to attend motorcycle rallies, races, and other motorcycle gatherings. At big annual gatherings such as Daytona and Sturgis, you will find Harley people working the crowds, talking to people, and finding out what peeves and pleases Harley owners and potential customers. They are paid to attend these events because of the importance the company places on staying connected with customers. Not surprisingly, Harley ranks among *Fortune* magazine's "100 Best Companies to Work For."

Pay Close Attention to the Aftermarket

One of the reasons for the company's turnaround in the 1980s was the recognition that there is a huge aftermarket for clothing and accessories and that the Harley-Davidson brand name has enormous appeal within this market. Harley-branded parts and accessories now are a major source of revenue for the company. The ideas for these accessories come from close associations with customers. In many cases, designers pick up on styling ideas that originated in the popular Harley culture.

The company also connects with the customizing world by offering factory-built custom motorcycles. This is a small side business for a major manufacturer; however, it serves to fuel the styling creativity within the organization and keeps the company in touch with the trend-setting end of the consumer spectrum.

Demonstrate Products and Debrief Customers

At rallies and through customer focus groups, Harley-Davidson frequently demonstrates new products and new product concepts and it solicits customer reaction. The feedback from these demonstrations helps designers to refine their concepts.

Assimilate and Document

Because of the range of activities that keeps Harley staff close to the customer, the company enjoys a rich sense of context for the lifestyles of their customers and a feeling of partnership with them in meeting their needs. Out of this contextual background, it is the function of marketing at Harley-Davidson to interpret and focus these needs into detailed design requirements. New

Model	Type	Quality	Style	Comfort	Low Price	Performance	Sport	Racing
FL	Touring	+++	+	+++				
Softail	Cruiser	+++	+++ Classic	+ Improved				
Dyna	Custom	+++	++	+	+			
Sportster	Entry	+	+ Race		+	++	+	?
XR-750	Dirt Racing					+++		+++
VT-1000	Track Racing		Sport			+++	++	+++
V-Rod*	Cruiser	+++	+++ Drag bike	++		+++	++	
Buell Blast*	Entry		Sport	+	+++			
Buell Firebolt*	Sport		Sport	+	+	++	++	++

*Recent model introductions

Legend
+ Good
++ Very good
+++ Excellent
☐ Unique in class

Figure A.1 Personal assessment of Harley-Davidson/Buell model strategy. (Courtesy of A. George.)

product designs, therefore, are backed by extensive market requirement specification documents that define the specific needs that the new designs are targeted to satisfy.

Market Strategy Result

Since the 1980s Harley-Davidson has successfully grown its product line to cover most motorcycle interests and lifestyles. Its bikes range from touring to sport and racing. Furthermore, the company has embraced the customization market and will produce stripped-down bikes intended for customization. In addition, with factory-built custom models, it competes directly with custom-bike shops and small manufacturers of "Harley-type" bikes.

Harley-Davidson has had a satisfied and loyal customer base, but in the 1990s the company feared that its customers might "age away." That is, it was not seeing younger customers in the market purchasing its motorcycles. One of the problems was that most of its product line was expensive, large, and heavy. The company realized that it needed some entry-level bikes and steps "upward." You can see the market strategy that emerged from this analysis. Figure A.1 is a personal assessment by Professor George of the Harley-Davidson/Buell product line with his subjective judgments of different attributes—quality, style, comfort, low price, performance, and application (sport or racing). In Professor George's scheme, a blank cell is a neutral rating, a single plus ("+") means "good," and triple pluses ("+++") mean "excellent." Some of the newer additions to the product line (V-Rod, Buell Blast, and Buell Firebolt) are highlighted together with their unique attributes. The Buell models were added to the product line specifically to attract entry-level customers.

The introduction of a new model is a major decision for the company and is preceded by extensive analysis as to how it fits into a matrix of customer needs, how it will support a long-term market strategy (such as attracting younger owners), and how it compares with the competition. To this end, Harley-Davidson studies each attribute that is important to its customers (performance, style, sound, etc.) and benchmarks its product line against competitors. The responsibility for every attribute related to quality is assigned to some individual within the company to ensure that it is not neglected as the design activity proceeds.

Case Study: Formula SAE Racing Competition

To illustrate the concept of continuous improvement and the concept of iterating by level, we describe Cornell University's experience in the Formula SAE racing competition.

Context and Background

The Formula SAE® race car competition is organized by SAE International (formerly the Society of Automotive Engineers). It is an international competition with entries from over 100 universities in the United States, Canada, Mexico, Japan, Great Britain, and more.

Each team must design, manufacture, develop, and race a new, 600cc, F-1-style car each year. Every team constructs one prototype car and uses it in competition events. In addition to creating the physical prototype, each team must present a cost analysis and manufacturing plan for the production of 1,000 cars. The target customer in the business plan is a race-car enthusiast who might buy such a car to compete on weekends in Sports Car Club of America (SCCA) autocross events. Many of the Formula SAE rules are patterned after the SCCA rules.

Teams are scored according to a point system with differing point values for the various events of the competition: endurance, acceleration, skid pad, fuel economy, autocross, design styling, cost and manufacturing plan, and marketing. Some of the events are physical events and competitions involving the vehicle itself. Some events are student presentations of business, marketing, and manufacturing plans. The engineering design contest is conducted using judges who are professional automotive designers.

The Cornell Formula SAE Record

Cornell University's team is made up of graduate (master of engineering) and undergraduate (bachelor of science) students from several engineering disciplines participating in the design. The Cornell Formula SAE team has been world champion six times since 1997 and nine times in the 20 years in which it has been entered. (Figure A.2 shows the 2005 winning car and Figure A.3 shows the 2005 winning team.)

Why has Cornell succeeded so many times in this competition? The reason is that the school uses a disciplined systems approach. Professor George describes the process this way:

> When we go to these competitions and look over the competition, we are really impressed with the ingenuity that has gone into the other designs. So many of the designs have some really innovative aspect to them….You can see what excited that design team. But

Figure A.2   Cornell SAE race car 2005. (Courtesy of A. George.)

those teams don't win. They don't win because they haven't taken a balanced approach. They haven't studied what it takes to win across the board. They haven't made the tough trade-off decisions. To win the Formula SAE competition we need a strong placing in all of the constituent events. This cannot be achieved without a balanced design. It is clear that we need systems engineering and team approaches to win!

A. George, 2008

Define the Problem

The first step in the systems approach is to define the problem. In this case, defining the problem takes a number of forms, as suggested by the following.

Figure A.3   Cornell 2005 team with SAE Foundation Cup. (Courtesy of A. George.)

Mission Statement

At the beginning of each year, the Cornell team leaders review their goals for the upcoming competition. At a mission statement level, they boil down to just four goals:

1. Win the competition this year.
2. Prepare to win the competition in future years.
3. Provide an excellent educational experience.
4. Have fun while accomplishing these goals.

Rule Analysis

From the beginning, the Cornell team takes a balanced view of the design process and focuses all decisions on the scorecard. The team members develop a thorough knowledge of the rules of the competition.

Optimization

Over the years, the Cornell team has developed mathematical models to understand the impact of a variety of design parameters on the overall possible score. This permits the team to determine quantitatively whether an investment in improving performance in one category is worth the cost or the performance degradation that it may cause in another category.

Benchmarking

At every competition, the team makes notes comparing its car's performance with that of the competition's cars. These notes form the basis for the design discussions at the beginning of the next year.

Competitive Intelligence

When studying the competition, the team is always on the lookout for features or functions that could enhance the design. As Professor George says, "We are not always the first to innovate but we are often very successful at incorporating new ideas."

Brainstorming

The senior team leaders encourage and employ brainstorming techniques to generate new design ideas. However, they also impose budgets and deadlines on new development efforts. If a new idea does not survive a design review, they cut the effort. This is a difficult discipline for student developers to accept: They have the passion of inventors for their ideas. However, the vehicle development schedule is tight and the senior team leaders cannot allow any portion of it to slip.

Corporate Memory

One source of ideas for features, functions, and solutions is the past. There is a great tendency for student teams unwittingly to reinvent past solutions. This is a waste. In many cases, they can make greater progress by consciously adapting subsystems and components that have been developed in earlier competitions. However, this requires a corporate memory. Everyone needs to be able to find past designs and performance reports to use as the basis for new development. Consequently, the Cornell team takes a disciplined approach to documenting and archiving designs and reports from the current year for use in future years.

Examples of Derived Requirements

From analysis of the competition rules and the study of scoring trade-offs, the team derives various functional and nonbehavioral requirements. Some examples of the derived requirements include:

- Car shall finish acceleration event in less than 3.9 seconds. Note that this does not specify how the car is to satisfy the requirement. This is a requirement derived from the "win this year's competition" goal.
- Car shall corner at higher than 1.4 *g*. This requirement will drive a number of important design decisions, but it was derived after a scoring trade-off analysis and determined to be critical to success.
- Car must finish the endurance event. This requirement simply recognizes that no part of the scoring competition can be ignored without sacrificing the overall objective.
- New ideas for future years' traction control enhancement shall be developed. This requirement was derived from the high-level goal "win the competition in future years." Traction control has been identified as a critical technology for improvement.
- Students must record learnings. This somewhat vague requirement traces back to two high-level goals: "provide educational experience" and "win in future years." First, for an experience to be educational, the student should be able to articulate what lesson was learned. Requiring a record to be kept of these lessons, or *learnings,* contributes in a tangible way to that high-level goal. Second, the importance of corporate memory in improving future designs has already been stressed. Recording the lessons learned in a corporate archive can assist future students to cover similar ground more quickly.

Requirements Traceability

Recall the concept of linked houses of quality in Chapter 9, "Iterate the Design Process." The hardware and software specifications for every component must be understood in terms of the requirements for the higher level module, subsystem, or interface to which it belongs. If requirements conflict, it may be necessary to trace the requirements all the way to the highest level. For example, consider the following trace of a low-level requirement all the way back to a mission goal.

The performance requirements of a pressure sensor may be tied to the requirements of a waste gate controller that requires continuous measurement of exhaust pressure. The waste gate controller, in turn, performs one of the many functions required for the turbocharger. The requirement for the turbocharger is driven by the need for engine power. Required engine power was calculated as part of a trade-off analysis (with car weight and tire traction) to meet the required acceleration target. The acceleration target was set on the basis of a requirement to "win the acceleration event." The "win the acceleration event" requirement was derived from the "win this year's competition" goal and the study of scoring trade-offs. Thus, if there is a problem with acquiring a pressure sensor that can meet its performance requirements, the team knows immediately all the higher-level requirements that are affected. These requirements then help the students to prioritize the problem.

The 2005 race car was designed with the main goal of moving weight to the rear of the car in order to increase rear wheel grip. The new weight distribution coupled with increased low end torque from our new KP35 Borg Warner turbo greatly increased our corner exit acceleration. Our custom designed oil pan with a dry sump system enabled the engine to be lowered an inch closer to the ground, thereby lowering the CG of the car.

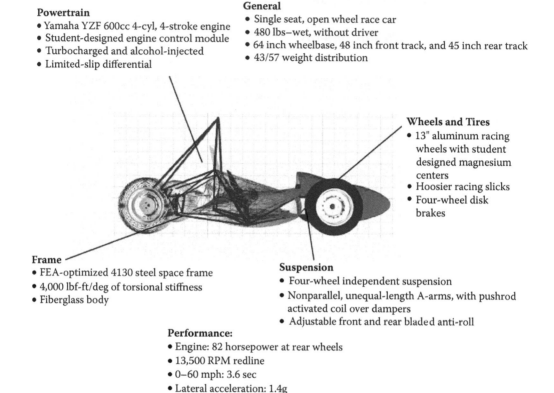

Powertrain
- Yamaha YZF 600cc 4-cyl, 4-stroke engine
- Student-designed engine control module
- Turbocharged and alcohol-injected
- Limited-slip differential

General
- Single seat, open wheel race car
- 480 lbs–wet, without driver
- 64 inch wheelbase, 48 inch front track, and 45 inch rear track
- 43/57 weight distribution

Wheels and Tires
- 13" aluminum racing wheels with student designed magnesium centers
- Hoosier racing slicks
- Four-wheel disk brakes

Frame
- FEA-optimized 4130 steel space frame
- 4,000 lbf-ft/deg of torsional stiffness
- Fiberglass body

Suspension
- Four-wheel independent suspension
- Nonparallel, unequal-length A-arms, with pushrod activated coil over dampers
- Adjustable front and rear bladed anti-roll

Performance:
- Engine: 82 horsepower at rear wheels
- 13,500 RPM redline
- 0–60 mph: 3.6 sec
- Lateral acceleration: 1.4g

Figure A.4   Features of the Cornell FSAE 2005 race car. (Courtesy of Al George.)

Architecture

Figure A.4 is an annotated product concept diagram that illustrates some of the major architectural and performance features of the 2005 Foundation Cup winner.

Numerous architectural decisions within the subsystems or at the interfaces must be made over the course of development. Each of these decisions is addressed by some form of trade study. The decisions with which the student teams must grapple include the following:

- How should the engine or electronics interface be defined?
- Should a separate or an integrated engine control computer be used?
- Should inboard or outboard brakes be used?
- For any component, should it be purchased, manufactured in-house, or modified from the previous year?
- When a metal part is cut, should manual or computer-controlled machining be used?

Architecture refers not only to the physical architecture of the vehicle but also to the operational architecture of the design, build, and test process. Here again, trade studies must be used to resolve questions such as

- How can the skills of team members be utilized best?
- Should driver training be conducted on a go-kart or on the car from the previous year's competition?

Management

Many of the students return to join the team year after year, progressing to higher levels of leadership each year. Management, budgeting, and scheduling are critical to success in the competition.

Resources

Senior team members are responsible for all aspects of the design organization, including financial management, capital investment, and personnel recruitment. In addition, it is an ongoing responsibility of the team to find and keep industrial sponsors. Senior team members are responsible for generating income, writing proposals, and obtaining sponsorships. They are also responsible for managing expenses and for budgeting the remaining funds.

Space is provided to the team by the university and funded by gifts from sponsors. The team is responsible for planning upgrades to the facility and for making improvements to the equipment.

Students who want to join the project team must go through an interview process. Senior team members are looking for particular needs to be fulfilled and must match applicants to needs. Occasionally, they must actively seek students with particular skills.

Planning

The Formula SAE team at Cornell is a small project team by many industrial standards, and it is subject to frequent turnover of participants because of the academic cycle. It relies on growing leadership and experience within the organization. It does not have the luxury of hiring talent from the outside. Therefore, the organizational chart must be adapted each year, as well as each semester, to fit the people who sign up for the project.

There is an extensive work breakdown structure of tasks that must be performed over the course of the design project. Team leaders are responsible for organizing their teams to accomplish these tasks and for maintaining a schedule of task deadlines and design review points.

Design review meetings are frequent and often focus on difficult trade-off decisions. Sometimes design ideas must be declared unworkable, even though their proponents beg for just a few more days to get them working. Efforts must be redirected into alternative paths with a higher probability of success. At some point, designs must be frozen to permit manufacturing to proceed, just as, at other points, design changes must be allowed.

Case Study Summary

The Cornell Formula SAE team is a microcosm of a design and engineering organization. Many of the issues faced by large design organizations are experienced by the students on the FSAE team. The Cornell team has a long history of success at winning the Formula SAE competition, and one of the critical factors in its success has been the use of systems engineering to establish balanced requirements, to conduct rigorous design reviews and extensive trade-off analyses, to manage teams, and to budget, schedule, and replan effectively.

References

Harley-Davidson Annual Report. 2004. Harley-Davidson Inc. Milwaukee, Wisconsin. http://www.harley-davidson.com/company/investor/ar/2004/_media/pdf/2004_HD_AR_full.pdf

Appendix B: Product and Service Development Challenges

The case studies in this appendix are hypothetical product and service development challenges. They are intended for classroom use as the basis for system design exercises. The "facts" and data of these cases are almost entirely fictional. They are intended to serve an educational purpose only. They are modeled after market requirement specification documents we have observed in industry.

Numerous individuals have contributed to the task of creating and editing these design challenges. Gavin Hurley, Mustafa Maqbool, Earl Valencia, and Nayan Dhanak created first drafts of these cases. Extensive editing came from the efforts of Albert George, Mason Peck, Rafaello D'Andrea, John Belina, and Lisa Dundon. We gratefully acknowledge their many contributions.

Table B.1 lists all the deliverables that would be expected in a complete system design exercise for any one of these challenges. A description of each deliverable can be found in the chapter referenced for that item. If you are conducting one of these design exercises as part of a course, note that your instructor will likely provide you with supplemental information for some of the deliverables.

Product Development Challenge: A Bathroom-Cleaning Robot

Preface

This is a fictional case study to serve as the basis for a product development educational exercise. Even though the inspiration for features of the case can be mapped in part to the current marketplace, no element of this case (products, companies, statistics, or regulations) should be considered factual.

Product Concept and Market Review

Company Profile/Industry Background

Helping Hands is a successful manufacturer of specialized robots for the hospitality industry. The company is a recent start-up and in 2005 launched the BathBot, which is a small robot designed to clean the bathtub in motel and hotel bathrooms. The technology breakthrough that Helping

Table B.1   Deliverables for Complete System Design

	Deliverable	Chapter
1	Mission statement	2
2	Annotated product concept sketch	2
3	Context matrix (or context diagram)	2
4	List of use cases	2
5	Use case behaviors	2
6	Functional requirements	2
7	Voice of the customer summarized into product objectives	2
8	Measures of effectiveness and data collection methods	3
9	Ranking and weighting of product objectives	3
10	Benchmarking of competitive products	3
11	House of quality	3
12	Technical performance measures and nonbehavioral requirements	3, 7
13	Customer value proposition	3
14	Alternative integrated product concepts	4
15	List of generic subsystems	4
16	Pugh analysis and recommended integrated product concept	5
17	Operational descriptions of subsystem interactions	6
18	Functional interrelationship matrix (or functional flow block diagram)	6
19	State change matrix (or state change diagram)	6
20	Interface, link, and interaction matrices	6
21	Initial bill of materials	6
22	Cost budget by subsystem	6
23	Reliability budget by subsystem	6
24	Test plan	7
25	Verification cross reference matrix	7
26	Failure modes and effects analysis	7
27	Work breakdown structure with durations and responsibilities	8
28	Table of task input–output relationships	8
29	Task precedence matrix	8
30	Project development schedule	8

Hands exploited was the ability to control a one-piece flexible arm accurately through the tension applied to surrounding cables. This eliminated the numerous joints and motors in conventional robotic arms. With a $3,000 price tag, the product was a resounding success from 2005 to 2007, selling mainly to hotel chains but also to independent hotel and motel owners who found that it saved their cleaning staff considerable time and effort.

The BathBot is designed to be light enough for the cleaning staff to carry from room to room. The cleaning attendant places the robot in the bathtub and then connects a water supply to the bath tap and a cord to an electrical outlet. The robot can move around on the floor of the bath. Waste water is dispelled through the bottom of the robot and rinsed down the drain.

The cleaning part of the robot consists of a flexible arm to which three end effectors can be attached: namely, a nozzle, a rotating cut-cloth mop (similar to those in automatic car wash facilities), and a blower. The nozzle first delivers a high-pressure spray of water and follows this by applying the cleaning solution. The cut-cloth mop then scrubs the surface after which the surface is rinsed using water through the nozzle again. Finally, a blower attachment is used to dry the surface. The robotic arm can detach, store, retrieve, and attach any of its three end effectors programmatically, without any human intervention.

The fact that no part of the BathBot comes in direct contact with a surface (the mop merely brushes against the bathtub) means that it is easy to configure the robot to adapt to different surfaces. This adaptation is not completely automatic, however. Although the robot does have some sensors, some information about the different sizes of bathtub must be programmed initially. The BathBot can store up to five such configurations, which the cleaning attendant can then activate.

The entire cleaning operation takes 3 minutes on average, during which time the cleaning attendant works on other areas of the bathroom or hotel room. This timesaver allows attendants to be more efficient and cover more bathrooms, which has been very well received by hotel managers.

The company's competitive advantage lay in being first to market, thanks to a highly skilled and innovative design and engineering team. Sales in 2005 exceeded $20 million. Helping Hands estimated that, at that point, only 10% of the potential market had purchased their product and thus the outlook for continuing growth was strong.

In addition, Helping Hands made a strategic decision to require consumers to purchase cleaning fluid cartridges from them. This greatly simplified the design problem because they did not have to test the effect on the robot of consumers refilling with different kinds of cleaning fluids. In addition, the refills were sold at a profit.

However, in 2008 Advanced Robotics, a large robotics manufacturer serving a variety of industries (automotive, semiconductor, and pharmaceutical), entered the market with a "copy cat" product, the RoboCleaner. Advanced Robotics was able to take advantage of its experience in the robotics industry and economies of scale to offer a similar product with a $2,500 price tag.

In addition, a second start-up firm, Hotel Robotics, entered the market. Hotel Robotics' product, the AutoCleaner, was also priced at $3,000. Apart from a more attractive appearance, the product was essentially the same as the BathBot, but less reliable. Helping Hands's market analysis concluded that the only reason AutoCleaner had any sales was from Hotel Robotics's industry contacts (the company was set up by former Hilton executives) and a highly effective marketing campaign.

Helping Hands responded immediately by pointing out to customers that it offers more focused customer service and greater reliability. Unfortunately, this was not enough to combat the pressure on price. In 2008, sales for Helping Hands' BathBot plummeted 65% to $7 million, whereas RoboCleaner had sales of more than $15 million. The company therefore faces being squeezed out of a market that it had created.

Market Opportunity

The management of Helping Hands has decided to face this challenge with bold action. It plans to design and build a robot, called the BathBot II, which will clean all three fixtures (tub, sink, and toilet) in a typical hotel bathroom. The cleaning attendant would be required to remove all towels and personal care products (shampoo, soap, etc.) from the bathroom and connect the robot to power, water, and waste disposal, but otherwise the robot would clean the fixtures unaided. Much like the BathBot, the BathBot II will require the various bathroom types to be programmed in.

Helping Hands will price the BathBot II at $3,500 a unit. The company expects that, because of the configuration required, this product will be economical only for hotels or motels where there is some uniformity in the layout of bathrooms. Thus, Helping Hands will focus its marketing efforts initially only on large hotel chains.

Market Conditions

Competitive Analysis

(a) Helping Hands
  - product name: BathBot II
  - capability: three-fixture bathroom (tub, toilet, and sink)
  - cleaning fluid: Helping Hands sells refill containers with cleaning fluid
  - maximum extension of robot arm: 1 m
  - average time to clean tub: 3 minutes
  - reliability: very reliable—MTTF (mean time to failure) is 600 hours of use
  - configuration memory: stores information on up to 35 types of bathrooms
  - weight: 10 kg (with full cleaning fluid container)
  - threats/claims: innovative, reliable product with excellent and focused customer service
  - justification: quality of design and after-sales service
  - drawback: product is more expensive both in initial purchase and in operating cost; customers are irritated by the need to purchase cleaning supplies from Helping Hands

(b) Advanced Robotics
  - product name: RoboCleaner
  - capability: tub only
  - cleaning fluid: removable 0.75 L container can be filled by using fluid of choice
  - maximum extension of robot arm: 1.2 m
  - average time to clean tub: 3.5 minutes
  - reliability: poor—MTTF is 200 hours of use
  - configuration memory: stores information on up to five different types of bathtubs
  - weight: 9.5 kg (when cleaning fluid container is full)
  - threats/claims: price—consumer can use his own cleaning fluid to refill
  - justification: price and more stable company to use as long-term supplier
  - drawback: market is not the company's main focus, meaning that customer service is neglected and innovation from Advanced Robotics is less likely

Helping Hands response:
  - Excel at innovation by developing the BathBot II.
  - Match cost to the customer of consumables (currently, customer must buy cleaning fluid from Helping Hands).

- Excel at customer service.
- Continue to offer a faster and more reliable product.

(c) Hotel Robotics
- product name: AutoCleaner
- capability: tub only
- cleaning fluid: refills shipped by Hotel Robotics
- maximum extension of robot arm: 1 m
- average time to clean tub: 4 minutes
- reliability: poor—MTTF is 250 hours of use
- configuration memory: stores information on up to 10 types of bathtubs
- weight: 12 kg (with full cleaning fluid container)
- threats/claims: design is more attractive; more effective marketing campaign
- justification: company executives have connections in the hospitality industry
- drawback: pricing is similar to the BathBot, but the product is less reliable

Helping Hands response:
- Draw attention to the inferiority of the product.

Customer Profile

The BathBot II will target hotel chains and large independent hotels with a large number of bathrooms of similar layout. The secondary target market will be independent hotel and motel owners with smaller premises.

Market Strategy

The market strategy is to focus on convenience in configuration, ease of use, a truly innovative product, high-quality customer service, low price, and low operating cost.

Market Size

There are almost 4.4 million hotel rooms in the United States, 74% of which are in hotels with more than 75 rooms (American Hotel & Lodging Association 2008). Helping Hands will target only these larger hotels. The company estimates that one BathBot II can serve approximately 20 rooms per day. This gives a total potential market of 165,000 BathBot IIs, or a $575 million market.

Table B.2 Actual and Projected Sales of Bathroom-Cleaning Robots

Annual sales ($)	Actual Sales				Projected Sales		
	2005	2006	2007	2008	2009	2010	2011
Helping Hands	20,000,000	22,000,000	24,200,000	15,000,000	40,000,000	44,000,000	48,400,000
Advanced Robotics	0	0	0	8,000,000	9,600,000	11,520,000	13,824,000
Hotel Robotics	0	0	0	3,000,000	3,900,000	5,070,000	6,591,000

Sales Projections

Helping Hands is very confident that its second innovation, the BathBot II, will be well received in the market. The company anticipates capturing 7% of the total market in 2009, giving a sales forecast of $40 million. Table B.2 and Figure B.1 illustrate the history of and projections for these sales.

Product Development Challenge: A Home-Health Monitoring and Trauma Alert System

Preface

This is a fictional case study to serve as the basis for a product development educational exercise. It is based in part on existing processes and capabilities but also features fictional projections of future technologies. Even though the inspiration for features of the case can be mapped in part to the current marketplace, no element of this case (products, services, companies, statistics, or regulations) should be considered factual.

Summary

The imperative to reduce costs in health-care systems is driving innovations in technology. In an attempt to switch from in-hospital care to remote patient monitoring, the current market is moving toward integrating wireless technology with medical instrumentation. In 2007, BA Biotech launched an all-in-one, hospital-based monitoring device for vital signs (blood pressure, temperature, pulse, and blood glucose) as well as for oxygen saturation and cardiac rhythm. To maintain its standing as a market leader, BA Biotech plans to launch HomeAlone, a product that will bring medical monitoring and trauma-alert care into the patient's home at an affordable cost.

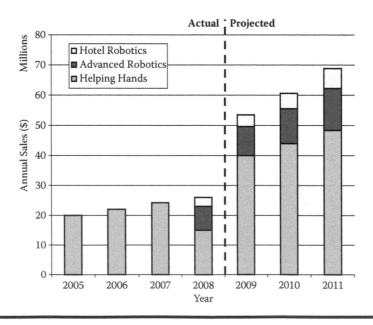

Figure B.1   Actual and projected sales for bathroom-cleaning robots.

HomeAlone's noninvasive remote monitoring will revolutionize the $1.5 trillion U.S. health-care industry. This product is designed to provide high reliability, high precision, and rapid response in case of emergency. On average, the payback period of investment will be less than 18 months, owing to insurance cuts and reduced visits to hospitals.

Product Concept and Market Review

Industry Background

Health-care systems around the globe are on the verge of a crisis, with financial stresses demanding an overall change. Technology has substantially enhanced the effectiveness of medical care by enabling the development of cures for previously fatal diseases and by advancing predictive and preventive medicine; however, this comes with an increasing price tag. An ever larger population and an increase in life expectancy have created a dilemma for health-care systems.

Innovations in monitoring devices over the past several decades have enabled increasingly rapid delivery of treatment to patients in need. As an additional result of these innovations, doctors are more aware of the medical history of their patients, and patients are experiencing greater mobility. The current market is moving toward integrating wireless technology with monitoring devices. This general application of wireless technology will increase the mobility of patients and caregivers, reduce costs, and greatly improve efficiency and profitability for the medical community because it will enable more patients to be treated in the home.

Even so, analysts believe that the $1.5 trillion U.S. health-care industry spends only 5% on information technology and even less on wireless technologies. Obstacles to increasing this percentage exist because of concerns over the reliability of these systems. Wireless transmission is vulnerable to signal interference, power loss, security breaches, and delays. The U.S. Food and Drug Administration (FDA) has issued stringent requirements on such devices, and these requirements have discouraged investment in this sector. However, technologies exist now that can satisfy FDA requirements, and the next decade is expected to see the integration of network technologies and health-care systems to improve patients' access to medical services at reduced costs.

According to FocalPoint Group, Inc. (2004), "The overall U.S. wireless data-networking and related servicing opportunity in the health-care sector will grow to over $7 billion by 2010, with the potential to be much higher given proper development."

Company Profile

The BA Biotech Corporation, in business for over 35 years, is widely recognized for its technologically advanced design and manufacturing capabilities and for its aggressive marketing of vital-sign-monitoring instruments. BA Biotech offers noninvasive monitoring solutions for blood pressure (NIBP), blood glucose, and oxygen saturation. It has maintained an average growth rate of 7% since 2005. Its recent innovations in monitoring devices have set a benchmark for other manufacturers, and its growth rate for 2009 (this year) is projected to exceed 10%.

In 2007, BA Biotech launched the first-generation Patient Probe RBC, which was an all-in-one monitoring device for vital signs (blood pressure, temperature, pulse, and blood glucose) as well as for oxygen saturation and cardiac rhythm. This model was launched in hospitals to enable caregivers to monitor several patients in close proximity. The Patient Probe RBC also made it easy for hospital personnel to maintain patient medical history (PMH) records. Within a short period of time, it became a best seller. However, competition is high; within the last year, over a dozen companies have offered similar products. Only two of these companies pose a serious challenge to BA Biotech's long-established reputation and market lead: MT Instruments and CAMA. MT Instruments offers low prices, but its products are less reliable. CAMA offers patient support and reliability.

Market Opportunity

To maintain its standing as a market leader, BA Biotech plans to launch a new product that integrates its all-in-one monitoring device with wireless technology. The proposed product will enable health-care providers to monitor patients continuously and remotely, thereby reducing the frequency of hospital visits. This product will be the latest in the line that includes Patient Probe RBC and will bear the name HomeAlone.

The HomeAlone device is designed to perform in-home monitoring of vital signs, oxygen saturation, and even blood glucose (noninvasive). It will send information to a central monitoring system (CMS) in the home for measurement, analyses, possible diagnoses, display, and notification of alerts to others, either at home or at call-and-dispatch centers. Lightweight and reliable, the HomeAlone device will be worn by the patient on a wristband (the WristBand). HomeAlone will enable certain patients to avoid prolonged hospitalizations.

BA Biotech's extensive research on the thermal energy generated by metabolic reactions in the human body is the basis for its design of this lightweight, noninvasive blood-glucose monitoring device. The HomeAlone WristBand will use special sensors to measure temperature and light characteristics in a patient's wrist with a contact thermometer, a radiation thermometer, and a multiwavelength reflective-dispersion photometer. The WristBand will enable the central monitoring system to monitor blood glucose.

Able to move around while wearing the WristBand, patients will stay informed about abnormal vital sign readings thanks to an alarm system in the CMS (max 85 dB). In addition, this alarm system will notify the user of any system malfunctions, such as signal failure or hardware or software problems. Readings will be taken by the WristBand at fixed intervals and transmitted to the central monitoring system. The sequence and time interval between each vital-sign reading will be set so that measurement of one vital sign does not affect measurement of the other: For example, body temperature will always be recorded before blood pressure is measured. The patient's doctor will prescribe the frequency of each vital-sign reading and the threshold limits for initiating alerts. However, to ensure the timely detection of patient trauma, the maximum interval between any set of readings will be 4 minutes.

For near-continuous monitoring, the patient with HomeAlone will be required to wear the WristBand at all times except when washing or showering. The WristBand will employ a rechargeable battery pack and will be connected to the central monitoring system via a wireless connection based on code division multiple access (CDMA) technology. This connection must consume minimal power and maintain signal integrity despite interferences. To further conserve the battery life of the WristBand, the signal-reception sensitivity of the central monitoring system will be high. The CMS will have two USB ports to support external diagnostics and data transfer.

The central monitoring system will construct and retain a PMH and a history of present illness (HPI). BA Biotech will offer patients the option of connecting their HomeAlone to a nearby BA Biotech unit (call-and-dispatch station) via a cellular connection and a backup phone line. (The patient's home will need to be within cellular range to benefit from this option.) In case of a trauma alert, this nearby call-and-dispatch station will first attempt to speak with the patient via the built-in microphone and speaker on the WristBand. If there is no response or the patient's condition is critical, the call-and-dispatch station will dispatch an ambulance. It will also transmit the patient's medical history to the receiving hospital. A door-locking mechanism requiring a security code will be provided for the main entrance to the house. The call-and-dispatch station will communicate the security code to the emergency personnel arriving at the patient's home.

Because two systems will be in place, BA Biotech is seeking separate FDA compliance for each system: one for the alarm system inside the house and the other for the call-and-dispatch station response system.

This product will be sold for a target price of $8,500. For most patients requiring continuous monitoring, the savings on in-patient care will justify this investment within 18 months. BA Biotech will offer several purchasing options for this product, and the company anticipates that medical-insurance companies will assist patients in the purchase of these units.

Anticipating the market demand for such a product, competitors MT Instruments (MT) and CAMA (Computer Assisted Monitoring of America) have announced plans to offer similar products. MT has established a reputation for offering alternatives for a number of expensive medical instruments, although it is unable to offer high-reliability products and lacks aggressive marketing capability. MT is usually after lower tier customers who cannot invest in expensive solutions, so its presence is not the primary threat to BA Biotech. MT Instruments' proposed product will have similar features to those of HomeAlone, and it will likely be priced at $6,500.

CAMA has long been in the business of providing high-tech testing equipment for research purposes. It pioneered some of the noninvasive medical instruments available on the market. CAMA's products are usually expensive; however, it offers excellent maintenance support and has cultivated loyal customers in the hospital industry. Its existence has always been a threat for BA Biotech. Its version of HomeAlone will likely be priced at $9,500.

Market Conditions

Competitive Analysis

Customers for in-home monitoring are primarily concerned with price, reliability, weight of wristband, and patient support. Neither HomeAlone nor the similar products by MT Instruments and CAMA are on the market at this time. However, based on data from previous products from these companies, the following hypothetical comparison can be used for market-planning purposes:

(a) BA Biotech Corporation
  - product: HomeAlone
  - weight: wristband weighs 100 g
  - reliability (mean time between failures): 3 years
  - price: competitive
  - warranty: 2 years (repair or replace with rebuilt unit)
  - memory: wristband has 5 Mb; central monitoring system has 20 Gb
  - display: wristband comes with an LCD display featuring alarm, speaker volume, and varying fonts and brightness
  - communications: network is based on CDMA technology
  - service: service representative will respond to service call within 24 hours
  - threats and weaknesses: known for delays in launching product and in offering upgrades; components are not readily available on the market and can be purchased only from BA Biotech

- strengths and opportunities: experience in producing similar products; long-standing market leader; competitive prices; high reliability; patient support

(b) MT Instruments
- product: Probe 2000
- weight: wristband weighs 200 g
- reliability (mean time between failures): 1 year
- price: $6,500
- warranty: 2 years (repair or replacement expected during warranty)
- memory: wristband has 5 Mb; central monitoring system has 25 Gb
- display: central monitoring system has an LED display featuring varying fonts but only low resolution and brightness
- communication: uses frequency modulation
- parts: readily available on the market; can be replaced by user
- threats and weaknesses: customers constitute only 8–9% of the market; MT has a reputation for low reliability
- strengths and opportunities: customers are always looking for cheaper alternatives and MT Instruments usually has something to offer; MT has many sales contacts in the market and the highest number of repair centers

BA Biotech's response to MT:
  • The HomeAlone product will weigh less, offer a better display, and have a longer battery life.
  • MT's products have a lower reliability rating, and its patients are likely to attempt repairs by themselves. In addition, signal interference in frequency modulation is likely in the Probe 2000 and therefore transmission errors may occur. Finally, the Probe 2000 will likely feature an unattractive LED-based display.

(c) Computer Aided Monitoring of America (CAMA)
- product: Patient Care Remote
- weight: wristband weighs 150 g
- Reliability (mean time between failures): 4 years
- price: $9,500
- warranty: 3 years (repair or replace with rebuilt unit)
- memory: wristband has 8 Mb; central monitoring system has 40 Gb
- display: central monitoring system has a high-resolution plasma display featuring varying fonts and brightness
- communication: uses a frequency-hopping CDMA (FH/CDMA); high security; interference is extremely unlikely
- parts: spare parts are readily available on the market
- service: service representative will respond within 12 hours; excellent patient support
- threats and weaknesses: low battery life; high cost
- strengths and opportunities: CAMA has been in the business of providing medical instruments longer than BA Biotech and exceeds BA Biotech in maintenance support, innovation, and reliability; CAMA will also be coming up with ECG upgrades within a short period of launching Patient Care Remote

BA Biotech's response to CAMA:
  • BA Biotech's products come close to CAMA's products in reliability, patient support, and innovation of product. BA Biotech's products are available at lower prices, and the HomeAlone wristband will weigh less than CAMA's product.

Table B.3 Annual Unit Sales Projections for HomeAlone

Year	2009	2010	2011	2012	2013
Annual unit sales projection	40,674	55,905	76,614	104,716	142,785

Patient Profile

The first-generation Patient Probe RBC was launched for hospitals and medical centers. However, HomeAlone targets the home-dwelling elderly, patients requiring monitoring, and remote medical units with limited staff. Although this product appears to be expensive for individuals to own, it has an estimated payback period of less than 18 months. Savings come about through reduced annual medical-insurance claims and reduced hospital visits. Each user is likely to use HomeAlone for at least 5 years. Patients above the age of 65 are vulnerable to medical problems late in the night. Some patients will require monitoring of vital signs with high frequency; therefore, these patients will have high-priority status and service contracts will vary accordingly.

Market Strategy

The health-care industry is a competitive industry, so alternative solutions are always available on the market. The competitive edge held by BA Biotech is in reliability, cost, and patient support. However, to maintain its market leadership, business goals should also include optimal patient mobility, continuous monitoring, personalized packages, and theft and tamper prevention. Resources have to be allocated accordingly to bring about a change in the industry. Patients should be able to gain real-time information and better care and to self-treat and improve their overall health condition outside the confines of a hospital or doctor's office. Technology should focus on innovations to improve patient productivity levels and to reduce overall costs for medical treatment. HomeAlone will increase sales for BA Biotech by implementing this market strategy.

Sales Projections

Investment. Launching HomeAlone requires resources for research, test design, manufacture, and marketing. BA Biotech already has a research and development department, and the expected investment for launching HomeAlone is $2.5 million.

Forecast. HomeAlone RBC is primarily targeted at patients who are above 65 years of age. Forecast unit sales are shown in Table B.3 and Figure B.2.

Product Development Challenge: A Night Vision System for Automobiles

Preface

This is a fictional case study to serve as the basis for a product development educational exercise. It is based in part on existing processes and capabilities but also features fictional projections of future technologies. Even though the inspiration for features of the case can be mapped in part to the current marketplace, no element of this case (products, services, companies, statistics, or regulations) should be considered factual.

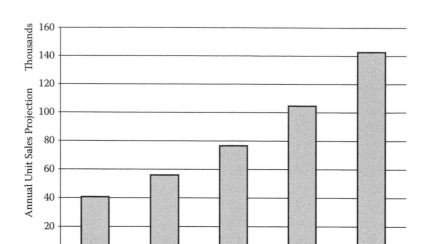

Figure B.2   Annual unit sales projections for HomeAlone.

Product Concept and Market Review

Industry Background and Company Profile

Nite-I Corporation has developed, manufactured, and supplied night vision equipment for the U.S. Army for over 15 years. The company is the acknowledged technical leader in infrared scanning and detection. The Army's strategic decision to build its capability to operate in darkness has ensured a steady stream of contracts for Nite-I. The company has recorded annual sales in excess of $800 million in recent years.

The night vision industry is currently poised to enter the civilian market. Two years ago, Terishimi, a Japanese electronics conglomerate, partnered with Mazota, an automobile manufacturer, to provide a night vision aid called Safe@Night that was built into the windshield of Mazota's luxury line of cars. This factory-installed option uses infrared scanners placed in the car's bumper to detect heat-emitting objects (pedestrians, animals, other cars) in the driver's field of vision. The car's windshield has an embedded display system that alerts the driver to the presence of these objects by marking them with a red dot.

This option was offered for an additional cost of $1,000. Customers responded very well, with over 70% of those purchasing a luxury Mazota opting for its inclusion. Mazota has announced that within 2 years the option will be offered on all models of car and included as standard in its luxury range.

Although the uptake of the option has been strong, Mazota's detection system does have significant flaws. The windshield refresh rate of 1 Hz often leaves drivers confused, as does the system's inability to distinguish between heat-emitting objects of different sizes. In addition, the display system is prone to failure; *Consumer Review* estimates that it has a mean lifetime of 3 years. Repair of the unit is expensive, often costing $500.

Market Opportunity

Nite-I's management is keen to tap into what it views as a very lucrative new market. The company has been testing a prototype of a similar product for Humvees with the Army and wishes to design a version for civilian use. It plans to partner initially with Mammoth Motors, a major

U.S. automobile manufacturer, to install such a product, called the NiteScreen, in Mammoth Motors cars.

Mammoth Motors wishes to follow Mazota's strategy of first offering an optional extra, costing around $1,000 in its luxury range of cars. To compete with Mazota, the company insists that the NiteScreen have a more frequent refresh rate and in some way distinguish between large and small objects.

The prototype of the NiteScreen being developed for the U.S. Army was required to be operable under extreme conditions and have superior reliability. In addition to this, Nite-I's skunkworks team fused together two existing technologies in order to create the patented ultrasonic-IR sensor array that has the higher accuracy in direction and motion needed for nighttime military missions. Nite-I's engineering team is confident that this can be continued in the civilian version, and this will give the company a significant competitive advantage over Mazota/Terishimi's product.

A third company offers a somewhat similar product. ScanTech, a spin-off from a university research group, offers the ThermoScanner, a unit that rests on the dashboard and displays information on its LCD screen, rather than on the windshield. The unit costs less ($600) and has had reasonable sales among electronics enthusiasts and other hobbyists, principally because it can be taken from the car and used in other contexts (night hunting, for example). However, Nite-I believes that the company does not have a chance of capturing the general market because drivers find it irritating to have the information on a separate screen. The ThermoScanner is also difficult to install and is not supported by any automobile manufacturer.

Market Conditions

Competitive Analysis

(a) Nite-I
   - product name: NiteScreen
   - detection system: fused ultrasonic infrared sensor array in bumper
   - display: dots displayed directly on windshield marking the object; size of dot corresponds to the size of the object
   - detection range: 60 ft (18.29 m)
   - refresh rate: 40 Hz
   - cost: $600 per unit
   - average lifetime: 7 years
   - repair cost: N/a
   - threats/claims: years of experience in sensor technology, especially in the defense industry
   - justification: reliability; robustness
   - drawback: unproven technology in the civilian market

(b) Terishimi
   - product name: Safe@Night
   - detection system: infrared sensor array in bumper
   - detection range: 50 ft (15.24 m)
   - refresh rate: 1 Hz
   - display: glass LCD—objects marked by red dot
   - cost: $1,000 per unit
   - average lifetime: 8 years
   - repair cost: $500

Table B.4 Actual and Projected Sales for Automotive Night Vision Systems

Annual sales ($)	Actual Sales		Projected Sales			
	2007	*2008*	*2009*	*2010*	*2011*	*2012*
Terishimi	18,000,000	19,000,000	10,740,000	7,606,000	8,096,200	8,554,300
ScanTech	0	800,000	1,040,000	1,352,000	1,757,600	2,284,880
Nite-I	0	0	10,000,000	15,000,000	16,500,000	18,150,000

- threats/claims: technology already in production
- justification: both Terishimi and Mazota are reputable, innovative companies with sub-stantial R&D budgets to iron out initial difficulties
- drawback: reliability—system can be confusing for drivers and thus dangerous

Nite-I response to Terishimi:
- Develop product that has superior technical performance (can distinguish objects by size, better refresh rate).
- Provide superior reliability.

(c) ScanTech
- product name: ThermoScanner
- detection system: portable infrared sensor array
- display: unit has its own 3 in. × 5 in. display rather than projecting on windshield
- detection range: 60 ft (18.29 m)
- refresh rate: 20 Hz
- cost: $600 per unit
- average lifetime: 6 years
- repair cost: N/A
- threats/claims: cheaper product; product failure does not require windshield repair
- justification: attractive to hobbyists and can be used in other contexts (hunting)
- drawback: difficult to install and distracting for drivers to use

Nite-I response to ScanTech:
- Partner with Mammoth Motors to offer a product that is attractive to the general market.

Customer Profile

The NiteScreen will target automobile consumers interested in buying the luxury line of Mammoth Motors—especially people who drive on poorly lit roads such as those in rural and suburban areas.

Market Strategy

A strategic partnership with Mammoth Motors will give Nite-I's NiteScreen significant access to the market. With the introduction of a different kind of sensor, this would provide the company with the necessary technology advantage. In addition to this, a focus on reliability, cost, accuracy, and detection speed (via refresh rates) will differentiate it from the competition.

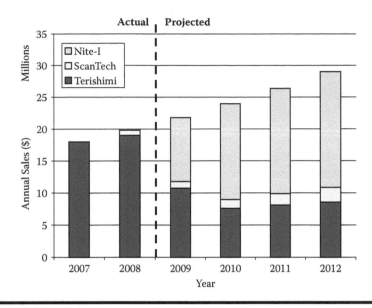

Figure B.3   Actual and projected sales for automotive night vision systems.

Market Size

Total sales of luxury vehicles in the United States topped the 1.6 million mark for the 2002 fiscal year.

Sales Projections

Mammoth Motors predicts to increase its market share by introducing the NiteScreen incorporated with its newest luxury model. Table B.4 and Figure B.3 show the resulting sales projections.

Service-Development Challenge: Internet-Based Meal Order and Delivery System

Preface

This is a fictional case study to serve as the basis for a service-development educational exercise. Even though the inspiration for features of the case can be mapped in part to the current marketplace, no element of this case (products, services, companies, statistics, or regulations) should be considered factual.

Product Concept and Market Review

Industry Background

The food-service industry in the form of fast-food or table-service restaurants is forecast to advance 4.4% in 2009 and to constitute 4% of the U.S. gross domestic product. The overall economic impact of such businesses is expected to exceed $1.2 trillion in 2009, including sales in related industries such as agriculture, transportation, and manufacturing. Between 1970 and 2009, restaurant and fast-food industry sales have posted a compound annual growth rate of 7.1%. This industry's share of the food dollar is estimated at 46.4%, up from 25% in 1955. This year, the nation's 878,000 restaurants should hit $440 billion in sales according to the National Dining Association's *2009 Dining Industry Forecast.*

Restaurants are the nation's largest private-sector employer, with one-third of all adults having worked in the restaurant industry at some time during their lives. Nine out of ten salaried employees at table-service restaurants started as hourly employees. The number of food-service managers is projected to increase 15% in the next 10 years.

The trend to eat out shows no sign of abating. Seven out of ten adults agree that there are more nutritious foods available to them in restaurants now than there were 5 years ago.

The Internet is changing the competitive environment for restaurants and fast-food outlets. A full 40% of consumers at restaurants and fast-food outlets have used e-mail or the Internet to find out information about establishments they have not patronized before.

Company Profile

Culinease Corporation is a food-distribution company delivering foodstuffs and partially prepared meal items to restaurants. It has been in business since 1992. By means of its excellent production and distribution system, broad product line, competitive pricing, and rapid-response deliveries, Culinease has been successful in establishing strong relationships with its customers, the table-service restaurants. As the restaurant business has grown, Culinease has increased its warehousing capacity, and the company is currently offering the highest variety of foodstuffs and meal items of any restaurant supplier in the country.

Although Culinease's sales have been growing an average of 10% per year since 1998, the future appears uncertain. Several competitors have recently entered the market and eroded Culinease's market share, currently at 18%. Two competitor corporations, TasteRight and BudgetShrinker, have successfully captured 15 and 14% of the market, respectively. TasteRight offers excellent customer support and reliability, while BudgetShrinker offers the lowest prices.

Culinease has eight production facilities and over two dozen warehouse facilities. It has over 50 suppliers and serves over 7,500 restaurants. In 2007, Culinease installed enterprise data management software in all of its facilities, which helped improve its overall efficiency. In an effort to extend its reach, Culinease has made several attempts to establish its presence on the Internet; however, its investments have not been fruitful.

Unfortunately for Culinease, most of its facilities are underutilized. The Culinease management has been looking at its options for expanding the business. In the past, Culinease has grown by evolving its product line. However, such opportunities are becoming rarer. Despite 9% growth in 2008, Culinease's market share dropped by 2%.

Market Opportunity

To maintain its standing as market leader, Culinease plans to enter the promising new arena of restaurant-meal home delivery by coordinating a meal-preparation and delivery service that will involve over 4,000 restaurants all over the United States. The name it has chosen to give this new business is NetMeals Tonight.

The initial system concept is for users to register on a Culinease Web site, where they will enter their postal zip code and then have access to all the possible menus available in their area. These menus will include both Culinease-brand meals and custom meals from the individual restaurants. Not all zip codes will be served. Some menu items will be available for immediate delivery as in the traditional takeout business. Other menu items, such as Peking duck, will need to be ordered at least 24 hours in advance to allow the Culinease logistics system time to complete the order.

Once customers place their orders, the Culinease meal-plan system will assign each order to a particular restaurant, and the local delivery system will be alerted to schedule both pickup and delivery. The delivery system will be operated like a taxi service, with vehicles and drivers serving multiple restaurants and consumers within a well-defined geographical region. In addition to offering consumers a wide variety of menu choices, the Culinease meal plan will also enable its users to customize meal preparation by noting their food allergies, favorite ingredients, and spice preferences. Users will be able to schedule and update meal orders up to a month in advance.

NetMeals Tonight will consist of two basic subsystems: the front end and the back end. The front end will deal with the customer interface to the system: the Web site and database and the ordering, payment, and user-tracking functions. The back end will deal with the restaurants and local delivery systems. The back-end system will interface with the existing Culinease system that manages all of the upstream operations (raw-material supply, production, warehousing, and distribution to restaurants).

The advantages of NetMeals Tonight from the individual restaurant's perspective will be to (a) increase its meal-preparation volume without having to increase its waitstaff, dining facility, or delivery staff; and (b) gain greater publicity for its unique menu. The advantages from the consumer's perspective will be a dining Web site that is available at all times, remembers individual preferences, and vastly simplifies the meal-planning process. For those who are interested, the Web site could enhance the menu-planning process with suggestions, nutrition calculations, and side journeys into such things as the history of particular recipes.

For Culinease, the system will put the company into direct contact with consumers and enable it to grow its brand of prepared meals and to accumulate a large database of consumer preferences and behaviors. The cost per delivered meal will be kept low because the system will make effective use of large Culinease kitchens, which enjoy economies of scale, and local restaurant kitchens, which can add last-minute meal customizations, such as fresh vegetables. By encouraging consumers to schedule orders days and weeks in advance, Culinease will have superior forecasting abilities that will enable it to optimize production schedules and to deliver prepared meals to restaurants in time to meet customer orders. Of course, a certain percentage of orders will be subject to cancellation and change at the last minute, so the system will require some safety stock.

With the implementation of NetMeals Tonight, Culinease forecasts that its growth rate will exceed 15% the first year and exceed 25% the next.

The two major competitors to Culinease, TasteRight and BudgetShrinker, are likely to enter this market themselves with similar offerings. The following competitive analysis is a projection of how their services are likely to compare, based on their current business practices.

Market Conditions

Competitive Analysis

The following is a brief comparison of the services provided by the competitors of Culinease, based on their anticipated features.

(a) Culinease Corporation
  - service name: NetMeals Tonight
  - delivery charges: moderate
  - items on the menu: 700+
  - on-time delivery record: excellent
  - delivered food quality: very good to excellent
  - Web site interface: superior consumer experience
  - threats and weaknesses: lack of experience in consumer market (but competitors have similar lack of experience)
  - strengths and opportunities: long-standing market leader; experienced in preparing basic meals and providing raw materials to restaurants; competitive (if not lowest) prices; high reliability; and customer support

(b) TasteRight Corporation
  - service name: Web Diner Deluxe, a name emphasizing its upscale nature
  - delivery charge: $8 on minimum order of $20; $10 or more for out-of-zone orders
  - restaurant database: 2,000 restaurants nationwide, typically high priced and gourmet style
  - items on the menu: 500+
  - on-time delivery record: excellent
  - delivered food quality: very good; TasteRight claims to preserve the original restaurant taste of the meal even after transport to the consumer
  - Web site interface: well designed, interesting, reasonably fast
  - threats and weaknesses: high delivery charge; limited variety of restaurants
  - strengths and opportunities: TasteRight has established a reputation with customers for on-time delivery

  Culinease's response to TasteRight:
  - Match quality at a lower price and emphasize value.

(c) BudgetShrinker Corporation
  - service name: Cyberfood 2U, a name emphasizing mass appeal
  - delivery charges: free delivery within zone
  - restaurant database: 3,000 restaurants, typically lower priced and family style
  - items on the menu: 250+
  - on-time delivery record: poor
  - delivered food quality: mediocre (frequently cold)
  - Web site interface: reasonably fast but not interesting
  - threats and weaknesses: lowest prices on meals and deliveries; cuts corners on quality; Web site look and feel say "cheap"
  - strengths and opportunities: established reputation among budget-conscious consumers for guaranteeing the lowest price

  Culinease's response to BudgetShrinker:
  - Emphasize quality in both back-end and front-end consumer experience.

Customer Profile

The target market for this new service will consist of singles, couples, and families who frequently eat out at table-service restaurants, but who would eat at home if the quality of food matched in-restaurant meals and if the planning, ordering, and delivery of meals were simple and reliable.

Table B.5 Projected Annual Sales by Culinease for NetMeals Tonight

	2009	*2010*	*2011*	*2012*	*2013*
Projected Annual Sales	$14,400,000	$50,400,000	$72,000,000	$93,600,000	$100,800,000

Market Strategy

Success in this new market will require excellence in both back-end and front-end systems. The front-end system must be both engaging to consumers and efficient at taking them through the order-entry process. The back-end process must be effective in transmitting the order throughout the supply chain to ensure rapid meal preparation and delivery as well as stock replenishment. Meal-preparation and delivery processes must be designed and performed so as to ensure consistent quality and exceptional food safety. Meal price and delivery charges must be within the range identified by competitive analysis.

Projections

Investment. No capital expenditures are required for food production or delivery. The new business will use the existing Culinease production and distribution system and the kitchens of Culinease's current customer base of restaurants. Delivery to consumers will be contracted to third-party delivery companies, such as local taxi services. Quality-of-service guarantees (such as "98% of deliveries shall be on time") will be negotiated in the contracts with these third parties. The major investment will be in the development of the front-end Internet-based menu browser and order-placement system. The basic system is projected to cost $1 million to develop and populate with data. Ongoing annual maintenance and support is expected to be 10% of that figure.

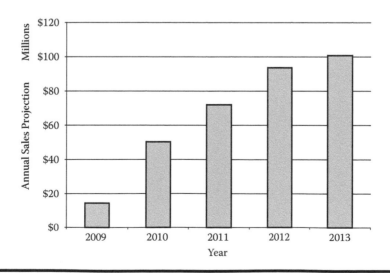

Figure B.4 Projected annual sales by Culinease for NetMeals Tonight.

Sales projections. Culinease estimates that a typical restaurant participating in the NetMeals Tonight system will generate $96,000 in sales of delivered meals per year. Of this, the sales revenue to Culinease will be 15%, or $14,400 per year. Over a 5-year period, Culinease will increase the number of restaurants participating from 1,000 in 2009 to 7,000 in 2013. Table B.5 and Figure B.4 project its anticipated growth in sales over these next 5 years.

References

American Hotel & Lodging Association. 2008. 2008 lodging industry profile. http://www.ahla.com
The FocalPoint Group. 2004. Wireless in healthcare: Executive summary. www.thefpgroup.com

Appendix C: Use Case Behaviors for Toy Catapult

In Chapter 2, only two use cases for the toy catapult are documented using behavior analysis. To be complete, we document the secondary use cases in this appendix. To begin, the following tables describe use case behavior:

- Table C.1. Behavioral Description of "Child Releases Armed Toy Near Face"
- Table C.2. Behavioral Description of "Child Aims Projectile at Eyes"
- Table C.3. Behavioral Description of "Child Uses Rodent as Projectile"
- Table C.4. Behavioral Description of "Child Uses Rodent as Projectile"
- Table C.5. Behavioral Description of "Child Plays with Toy Repeatedly"
- Table C.6. Behavioral Description of "Child Drops or Throws Toy"

After completing these behavioral descriptions of secondary use cases, we use the affinity process to summarize the secondary functional requirements and we expand the context matrix to reveal the aspects of system interaction that were uncovered through this analysis:

- Table C.7. Affinity Table for Secondary Functional Requirements
- Table C.8. Detailed Context Matrix for Toy Catapult after Secondary Use Cases

Finally, we simplify the context matrix by eliminating the internal entities that were used to describe the behavior:

- Table C.9. Summary Context Matrix for Toy Catapult after Secondary Use Cases

Table C.1 Behavioral Description of "Child Releases Armed Toy Near Face"

Child Releases Armed Toy Near Face of Self or of Another Child		
Initial conditions		
1.	The toy is in the armed state.	
2.		
Operator (parent and child)	**System (toy)**	**Projectile**
The parent acquires the toy.		
	The system shall notify parent of the dangers and suggest the appropriate age of child to use the system.	
The parent prevents use of toy by underage or untrained child. Yes **No**		
The child places armed toy near face of self or the face of another child.		
The child triggers the release.		
	The moving parts of the system shall be prevented from striking the face of a child with sufficient force to cause a bruise.	
Ending conditions		
1.	The toy is in the unarmed state.	
2.	The child is not seriously injured.	

Notes:

1. The toy can communicate dangers and suggest the appropriate age through its packaging, labeling, and instruction set, and through the general appearance of the toy.
2. It might be possible to design the toy so that it is impossible for the child's face to come in contact with a rapidly moving part.
3. Issue: For verification purposes, a substitute for a child's face must be found. A fresh apple may be an acceptable substitute to test for bruising.

Table C.2   Behavioral Description of "Child Aims Projectile at Eyes"

Child Aims Projectile at Eyes of Self or of Another Child		
Initial conditions		
1.	The toy is armed and loaded with projectile.	
2.	The projectile is pointed.	
Operator (parent and child)	**System (toy)**	**Projectile**
The parent acquires the toy.		
	The system shall notify parent of the inherent risk to eyesight and suggest the appropriate age of child to use the system.	
The parent prevents use of toy by underage or untrained child. Yes No		
The child aims projectile at eyes of self or of another child.		
The child triggers the release.		
	The system shall eject contents of the receptacle.	
	The system shall not launch a blunt projectile with sufficient force to penetrate a child's eye.	
		The projectile strikes the target without sufficient force to penetrate the eye.
Ending conditions		
1.	The toy is in the unarmed state.	
2.	The projectile does not penetrate the child's eye.	

(Continued)

Table C.2 Behavioral Description of "Child Aims Projectile at Eyes" (Continued)

Child Aims Projectile at Eyes of Self or of Another Child

Notes:

1. The terms "blunt" and "sharp" are imprecise.
2. Risk: It appears to be impossible to protect the child from all possible projectiles. For example, the use of multiple projectiles such as ball bearings or BBs increases the chance of an accurate and injurious strike.
3. Issue: For verification purposes, a list of acceptable but potentially dangerous projectiles from the child's domain must be identified. For example, the list should include crayons.
4. Issue: For verification purposes, a substitute for a child's eye must be found. A hard-boiled, unshelled egg may be an acceptable substitute for test purposes.

Table C.3 Behavioral Description of "Child Uses Rodent as Projectile"

Child Uses Pet Rodent as Projectile (1)		
Initial conditions		
1.	The toy is armed.	
2.	The pet rodent is agitated.	
Operator (parent and child)	**System (toy)**	**Pet rodent**
The parent acquires the toy.		
	The system shall notify parent of the dangers and suggest the appropriate age of child to use the system.	
The parent prevents use of toy by underage or untrained child. Yes **No**		
The child places pet rodent in receptacle.		
		The rodent attempts to escape. **Yes** No

Table C.3 Behavioral Description of "Child Uses Rodent as Projectile" (Continued)

	Child Uses Pet Rodent as Projectile (1)	
	The system shall enable escape of the rodent.	
		The rodent escapes.
Ending conditions		
1.	The toy is in the armed state.	
2.	The rodent has escaped the toy.	

Notes:

1. Issue: For verification purposes, the best test would include the use of a live rodent. The test procedure must guard against accidental release of the launch mechanism.

Table C.4 Behavioral Description of "Child Uses Rodent as Projectile"

	Child Uses Pet Rodent as Projectile (2)	
Initial conditions		
1.	The toy is armed.	
2.	The pet rodent is calm or asleep.	

Operator (parent and child)	System (toy)	Pet rodent
The parent acquires the toy.		
	The system shall notify parent of the inherent risk to pets and suggest the appropriate age of child to use the system.	
The parent prevents use of toy by underage or untrained child. Yes **No**		
The child places pet rodent in receptacle.		
		The rodent attempts to escape. Yes **No**

(Continued)

Table C.4   Behavioral Description of "Child Uses Rodent as Projectile" (Continued)

Child Uses Pet Rodent as Projectile (2)		
The child triggers the release.		
	The system shall eject the contents of the receptacle.	
	The system shall not launch a rodent with sufficient force to injure the rodent when it lands.	
		The rodent flies through the air and lands some distance away.
Ending conditions		
1.	The toy is in the unarmed state.	
2.	The rodent is not injured.	

Notes:

1. Risk: The toy is expected to have sufficient size and force to launch a pet rodent. The risk of injury to the rodent is therefore unavoidable.
2. Issue: For verification purposes, a substitute for a pet rodent must be found. A piece of peeled banana may be an acceptable substitute for test purposes to test for bruises.

Table C.5   Behavioral Description of "Child Plays with Toy Repeatedly"

Child Plays with Toy Repeatedly		
Initial conditions		
1.	The system is in the unloaded state.	
2.	The toy is new.	
Operator (child)	**System (toy)**	**Projectile**
The child pushes the receptacle into position.		
	The system shall detect the receptacle in proper position.	
	The system shall secure the receptacle in position.	
The child triggers the release.		

Table C.5   Behavioral Description of "Child Plays with Toy Repeatedly" (Continued)

	Child Plays with Toy Repeatedly	
	The system shall detect the command to release from the child.	
	The system shall eject the contents of the receptacle.	
The child repeats the cycle many times.		
	The system shall successfully complete each cycle of use.	
Ending conditions		
1.	The toy is in the unloaded state.	
2.	The toy successfully completed each cycle.	

Notes:

1. Issue: Need to resolve how many cycles the toy must survive.
2. The choice of materials has a strong impact on the durability of the toy.

Table C.6   Behavioral Description of "Child Drops or Throws Toy"

	Child Drops or Throws the Toy	
Initial conditions		
1.	The system is in either the armed or unarmed state.	
Operator (child)	**System (toy)**	**Projectile**
The child drops or throws the toy against a hard surface.		
	The system shall survive a collision with a hard surface in working order.	
Ending conditions		
1.	The toy is in working order (that is, the primary use case behaviors can still be performed).	

Notes:

1. Issue: The force with which a child can throw the toy increases with the age of the child. We need to identify the maximum force the toy will experience under normal conditions.
2. The design of the body of the toy has the most impact on this requirement.

Table C.7 Affinity Table for Secondary Functional Requirements

The System shall notify the parent of the dangers and inherent risks of the system.	The system shall suggest to the parent the appropriate age of the child to use the system.	The moving parts of the system shall be prevented from striking the face of a child with sufficient force to cause a bruise.	The system shall not launch projectiles from the child's domain with sufficient force to cause damage to the child or the projectile.	The system shall enable the escape of a pet rodent from the receptacle.	The system shall successfully complete each cycle of use for many cycles.	The system shall survive in working order from a collision with a hard surface.	Verification of all requirements shall be conducted in such a way as to preserve the health of human and animal subjects.
The system shall notify the parent of the dangers and suggest the appropriate age of child to use the system.	The system shall notify the parent of the dangers and suggest the appropriate age of child to use the system.	The moving parts of the system shall be prevented from striking the face of a child with sufficient force to cause a bruise.	The system shall not launch a blunt projectile with sufficient force to penetrate a child's eye.	The system shall enable the escape of the rodent.	The system shall successfully complete each cycle of use	The system shall survive a collision with a hard surface in working order.	Issue: For verification purposes, a substitute for a child's face must be found. A fresh apple may be an acceptable substitute to test for bruising.

The system shall notify the parent of the inherent risk to eyesight and suggest the appropriate age of child to use the system.	The system shall notify the parent of the inherent risk to eyesight and suggest the appropriate age of child to use the system.		Issue: For verification purposes, a substitute for a child's eye must be found. A hard-boiled, unshelled egg may be an acceptable substitute for test purposes.
	The system shall not launch a rodent with sufficient force to injure the rodent when it lands.		Issue: For verification purposes, the best test would include the use of a live rodent. The test procedure must guard against accidental release of the launch mechanism.
			Issue: For verification purposes, a substitute for a pet rodent must be found. A piece of peeled banana may be an acceptable substitute for test purposes to test for bruises.

Table C.8 Detailed Context Matrix for Toy Catapult after Secondary Use Cases

Is-Related to	Child	Parent	Toy	Receptacle	Lock	Spring	Release	Pkg.	Body	Projectile	Toy Train	Pet Rodent	Hard Surface
Child	aims toy at self or other		retrieves and plays with repeatedly	pushes into position		stores energy in	triggers			places in receptacle	anticipates triggering event from	uses as projectile	throws toy against
Parent	teaches, entertains, and trains in safety procedures		stores								aligns trajectory with toy release mechanism		
Toy	attracts, amuses, and does not harm			consists of	consists of	consists of	consists of	consists of	consists of			launches	
Receptacle	does not harm		is part of							holds and launches		does not harm	
Lock			is part of	secures in position		withstands force of	arms						
Spring			is part of	powers launch of									
Release			is part of		releases								

Packaging, labeling, and appearance	appeals to	informs of dangers and identifies appropriate child age	is part of						
Body			is part of	protects	protects	protects			survives impact with
Projectile	amuses and does not harm					triggers copy 2 of			
Toy train	amuses					triggers			
Pet rodent	amuses			escapes from or survives launch from					
Hard surface									

Table C.9 Summary Context Matrix for Toy Catapult after Secondary Use Cases

Is Related to	Child	Parent	Toy	Projectile	Other Moving Toys	Small Pets	Play Surfaces
Child	inappropriately aims toy at self or other		retrieves and plays with repeatedly; arms and loads; triggers release of	places in toy receptacle	anticipates triggering event from	inappropriately uses as projectile	drops or throws toy against
Parent	teaches, entertains, and trains in safety procedures		stores		aligns trajectory with toy release mechanism		
Toy	amuses but does not harm	warns of dangers and suggests appropriate age of child		holds and launches		does not harm	survives impact with
Projectile	amuses but does not harm						
Other moving toys	amuses		trigger				
Small pets			escape from or survive launch from				
Play surfaces							

Index